DAM THAT RIVER!

Ecology and Mormon Settlement in the Little Colorado River Basin

William S. Abruzzi
Penn State University

UNIVERSITY
PRESS OF
AMERICA

Lanham • New York • London

Copyright © 1993 by
University Press of America®, Inc.
4720 Boston Way
Lanham, Maryland 20706

3 Henrietta Street
London WC2E 8LU England

All rights reserved
Printed in the United States of America
British Cataloging in Publication Information Available

Library of Congress Cataloging-in-Publication Data
Abruzzi, William S.
Dam that river! : ecology and Mormon settlement in the Little
  Colorado River Basin / William S. Abruzzi.
      p.   cm.
Includes bibliographical references and index.
1. Mormons—Colonization—Little Colorado River Region (N.M.
and Ariz.)—History. 2. Ecology—Little Colorado River Region
  (N.M. and Colo.) 3. Land settlement—Little Colorado River
Region (N.M. and Colo.)—History. 4. Little Colorado River Region
        (N.M. and Ariz.)—History. I. Title.
  F817.L5A25   1993   979.1'33—dc20   93–17051 CIP

ISBN 0–8191–9126–4 (cloth : alk. paper)

 The paper used in this publication meets the minimum requirements of American National Standard for Information Sciences—Permanence of Paper for Printed Library Materials, ANSI Z39.48–1984.

This book is dedicated to my children,
Matt, Geof, Emily and Brian,
and to my father-in-law
William H. Reith, Sr.

Acknowledgements

Many people contributed to the completion of this manuscript. Brian Foster deserves special consideration for his critical support and guidance. Special acknowledgement is also made to the late Fred Plog for introducing me to the research possibilities offered by the Little Colorado River Mormon communities and for his years of professional and personal support. Appreciation is also expressed to Mike Little, Hal Koster and John Frazier for their help during my research and my preparation of the original manuscript. I would like to especially thank Mark Leone for his generous and unfailing support of my research and for his valuable suggestions on revising the original manuscript. Gratitude is also expressed to Marvin Harris for the efforts he has made on my behalf, on this and on other projects. Charles Peterson has also been helpful in sharing his valuable insights on the Mormon settlements investigated.

I would like to express my sincere appreciation to the many people of the Little Colorado River Basin who helped me during my research. It would be impossible to name all the people who shared their time and knowledge with me. However, special appreciation goes to the late Al LeVine, as well as to Joe Rodriguez, Vada Carlson, Randolph Randall, Delbert Hanson, Rudy Schnabel and Doug Immekus. Appreciation is also expressed to those individuals associated with libraries at the Historical Department of the Church of Jesus Christ of Latter-Day Saints, Arizona State University, the Arizona Department of Libraries and Archives, Northern Arizona University, the University of Arizona, The Arizona Pioneer Historical Museum, the University of Utah, and Brigham Young University for the enthusiastic help I consistently received.

I would like to especially thank Mary Reith for the dedication and hard work she put into the figures and maps included here and to Glen Kreider, Lois Seitz, Nancy Evans, Ilene Karp and Joanna Hill for their help in preparing the final manuscript for publication. And, finally, I would like to thank Arlene Taub for the special love and support she has consistently provided me that does not show up directly in the final manuscript.

This book is based on research supported by the National Science Foundation under Grant No. BNS 78 21489, and I wish to express my appreciation to NSF for its support. Invaluable financial and nonfinancial support was also provided by the Department of Anthropology at SUNY Binghamton, by Penn State University, and by both the Department of Anthropology and the Ogontz Campus of Penn State. I would like to express my appreciation to them as well.

Contents

Acknowledgement		v
List of Figures		viii
List of Maps		viii
List of Tables		viii
Preface		xi
1.	Introduction	1
2.	Colonizing the Little Colorado River Basin	19
3.	The Evolution of Ecological Communities	55
4.	The Little Colorado River Basin	79
5.	Dam Construction	121
6.	Exploiting Environmental Diversity	145
7.	External Impacts on the Settlement Process	165
8.	Conclusion	193
	References Cited	209
	Index	221

FIGURES

3.1	The Ecological Model	57
4.1	Average Monthly Runoff: Little Colorado River at Holbrook	97
4.2	Annual Runoff: Little Colorado River at Holbrook	98

MAPS

2.1	Little Colorado River Basin: Mormon Settlements	22
4.1	Little Colorado River Basin: Physical Features	80
4.2	Little Colorado River Basin: Plant Communities	87
4.3	Little Colorado River Basin: Sediment Yield	99
7.1	Little Colorado River Basin: Non-Mormon Factors	167

TABLES

2.1	Population of the Little Colorado Settlements, 1877-1880	28
2.2	Population by Ward, September 1879 and March 1880	28
2.3	Total Tithing by Ward, 1887-1905	37
2.4	Tithing by Field Crops (Grain and Hay) by Ward, 1887-1900	38
2.5	Tithing by Livestock by Ward, 1887-1900	39
2.6	Population by Ward, 1876-1905	40
2.7	Per Capita Tithing by Ward, 1887-1905	41
2.8	Number of Occupations Listed in the 1900 Census by Town	42
2.9	Establishment Dates for Church Organizations by Ward	43
2.10	Number of Businesses Listed for 1905-1906 by Town	44
2.11	Rank-Order of Settlements by Specific Criteria	45
4.1	Elevation and Precipitation for Selected Location in the Little Colorado River Basin	82

4.2	Relative Contribution of Winter and Summer Precipitation at Selected Locations in the Little Colorado River Basin	83
4.3	Mean Number of Frost-Free Days Annually	85
4.4	Area and Relative Proportion of the Four Principal Plant Communities in the Little Colorado River Basin	88
4.5	Average Estimated Yields of Principal Crops Grown on Selected Soils Under Prevailing Management Conditions	90
4.6	Monthly Distribution of Diversion from Daggs Reservoir, 1946	94
4.7	Snowpack Depth on Mount Baldy, 1950-1964	96
4.8	Monthly Discharge of the Little Colorado River at Holbrook, Arizona, March 1905-April 1907	100
4.9	Suspended Sediment Concentration in the Little Colorado River at Cameron, Arizona, 1969	102
4.10	Chemical Concentrations at Selected Sites along the Lower Valley of the Little Colorado River	103
4.11	Monthly Discharge of Silver Creek into Daggs Reservoir, 1942-1946	105
4.12	Stream Flow Summary for Silver Creek near Woodruff, 1929-1946	106
4.13	Annual Discharge at Selected Locations in the Little Colorado River Basin, 1930-1944	107
4.14	Variation in Monthly Discharge in Silver Creek, 1942-1944	109
4.15	Water Quality in Silver Creek at Woodruff and Daggs Reservoir	110
4.16	Number of Cattle and Sheep Registered in Apache County, 1916-1925 and 1958-1967	112
5.1	Dam Losses among Specific Settlements in the Little Colorado River Basin	123
6.1	Labor Expended at St. Joseph, 1879	147
6.2	Percent of the Population under 8 Years of Age among Settlements in the Lower Valley of the Little Colorado River, 1877-1886	149
6.3	Men, Women, Boys and Girls in the Lower Valley Settlements, September 1878	150
6.4	Mean Family Size at St. Joseph and Snowflake, Arizona, 1892-1897	152

7.1	Character and Amount of Freight Shipped from Holbrook and Winslow, Arizona on the Atlantic and Pacific Railroad Line during 1885, 1888 and 1889	169
7.2	Character and Amount of Freight Delivered to Holbrook and Winslow, Arizona on the Atlantic and Pacific Railroad Line during 1885, 1888 and 1889	170
8.1	Rank-Order of Mormon Settlements by Indices of Population, Productivity, Stability and Diversity, 1887-1905	200

Preface

This book applies an explicit ecological model to explain successful Mormon colonization of the Little Colorado River Basin in northeastern Arizona. The model employed is an adaptation of the general model developed by plant and animal ecologists to account for the evolution of complex ecological communities. The manuscript developed as a result of a general dissatisfaction with previous analyses of Mormon settlement in this region and elsewhere and of a desire to assess the applicability of general ecological concepts and principles to human ecology. The manuscript eschews traditional idealist explanations for successful Mormon colonization and adopts instead a detailed materialist analysis of the settlement process. In addition, the manuscript applies ecological theory explicitly and deductively to explain not just the overall success of the colonization effort but also several specific empirical developments associated with the settlement process.

    The research focuses on about a thirty-year time period between 1873 and 1905. There were several reasons for limiting the investigation to these years. First, the manuscript is offered in comparison with previous explanations of the settlement process which focus on this specific time period. Second, by 1900 all Mormon towns in the region had been established and no further settlements were abandoned. In addition, the extensive immigration and emigration which characterized the previous years had declined, and significant developments related to the establishment of viable agricultural settlements in this demanding and variable environment ceased. Third, extending the analysis beyond the settlement period would have introduced unnecessary additional information which would have diluted the clear problem focus of the research. Fourth, by focusing clearly on the settlement period the manuscript demonstrates that, if properly formulated, tests of evolutionary principles can be performed on data from relatively limited time periods. Limiting the analysis to the settlement period also permits a full consideration of the details of this local historical event, allowing consideration of the full complexity of historical developments and their relation to general theoretical principles. Consequently, the manuscript contains numerous, detailed footnotes which elaborate on specific points while preserving the general flow of the analysis.

CHAPTER 1

Introduction

In an effort to escape recurring persecution during the nineteenth century, members of the Church of Jesus Christ of Latter-Day Saints (Mormons) emigrated to the mountains of the American West and established the nucleus of a new Zion in the Great Salt Lake Valley. Shortly after founding Salt Lake City in 1847, the Mormon Church organized and subsidized the westward migration of thousands of converts from the eastern United States and western Europe (see Stegner 1942, 1964; Arrington 1958:97-108). As a result of both population pressure and the desire to preserve Mormon security and autonomy in the face of the expanding non-Mormon American frontier, Church leaders in Salt Lake City initiated an extensive colonization program. Under their direction, over 500 hundred Mormon agricultural settlements had been established by 1900 in arable valleys throughout the Mountain West from the Canadian border region south to northern Mexico and westward to the Pacific Ocean. The Little Colorado River Basin in northeastern Arizona was the focus of but one of many Mormon colonizing efforts undertaken during the last quarter of the nineteenth century.

Following a failed attempt at colonization in 1873, five hundred Mormon pioneers established four colonies along the lower valley of the Little Colorado River in March of 1876. During the next several years, additional settlements were founded along the upper Little Colorado River, along Silver Creek, its principal tributary, and in isolated mountain valleys further south (see Map 1). Populating this remote and largely barren region with farming settlements was considered integral to the Church's larger geopolitical strategy. As with Mormon colonization elsewhere, the goal was to occupy every arable valley in this arid river basin with self-sufficient agricultural settlements that were socially, economically and politically aloof from encroaching non-Mormon influences in order to minimize growing non-Mormon settlement throughout the Church's expanding geographical domain. The Little Colorado Settlements also provided a temporary haven for many Utah polygamists fleeing federal prosecution by moving to Mormon settlements in northern Mexico (see Hardy 1969).

Mormon colonization of the Little Colorado River Basin was a success. By the close of the nineteenth century, nearly 20 Mormon farming communities existed throughout the region. While none of these towns ever achieved complete isolation from surrounding non-Mormon influences or independence from external resources, a fair degree of local self-sufficiency was, in fact, attained. Significantly, farming throughout the basin remained primarily a Mormon accomplishment, with Mormon settlements succeeding where others had failed. Local non-Mormon settlement was confined largely to towns established along the Atlantic and Pacific

Railroad and at other locations along the region's perimeter (see Map 5). Moreover, most Mormon towns retained their distinct character and identity until rapid economic development after 1960 completely transformed the region by increasing non-Mormon immigration and by converting local communities into economic satellites of the expanding Arizona regional political economy (see Abruzzi 1985).

Although the overall colonization effort was successful, establishing and maintaining viable agricultural settlements in this arid and climatically variable region proved exceedingly difficult. It was eventually achieved only at considerable personal and community expense and through the judicious application of substantial Church resources. Recurring floods, droughts, early frosts, dam failures and other calamities caused by climatic variability severely tested the adaptive capacity of these early pioneers. At the same time, spatial differences in climate, soil quality, length of the growing season, availability of suitable water for irrigation and other natural environmental characteristics yielded significant local differences in community development.

While some settlements grew to several hundred persons by the close of the nineteenth century, others contained barely more than a few families. Still others were abandoned within a few years of their founding, despite substantial Church assistance and a cooperative regional colonizing effort. Agricultural productivity also varied sharply between towns, as did the complexity of economic, social and political organization. With respect, therefore, to three fundamental indices of community development--population size, economic productivity and functional diversity--some settlements were clearly more successful than others. In other words, while most settlements in the basin remained economically marginal throughout the nineteenth century, some more closely approximated the nineteenth century Mormon cultural ideal of prosperous, functionally diverse farming communities than did others.

The eventual success of Mormon settlement in the Little Colorado River Basin was due in large part to the development of a system of resource redistribution among individual towns which mitigated the negative consequences of local environmental variability. The region's pronounced spatial diversity (see Chapter 4) provided a unique opportunity for early settlers to overcome local environmental limitations by integrating the productivities of several habitats into a single resource-flow system. However, while early Mormon settlers quickly recognized the adaptive advantage of exploiting and redistributing surplus resources among diverse local habitats, their efforts to establish a viable, multi-habitat resource redistribution system were not equally successful. The Little Colorado Mormon settlements developed two distinct multi-habitat resource redistribution systems during the nineteenth century (see Chapter 6). The first consisted of several productive enterprises jointly operated by the early settlements in the lower valley of the Little Colorado River. These enterprises, which were situated at higher elevations to the south, provided resources which could not be efficiently produced in the lower valley towns themselves. They, therefore, supplemented the variable farming productivity achieved along the lower Little Colorado River Valley. However, this system proved ineffective and was eventually abandoned. The second system of resource redistribution was based on the circulation of tithing resources. Tithing, most of which was paid in kind, was collected and stored by local Church leaders and given to those in need. Through tithing redistribution, individuals and towns which suffered poor harvests were able to gain access to surplus resources produced elsewhere in the basin. The system of tithing redistribution succeeded where the conjoint enterprises had failed and proved to be a critical factor underlying successful Mormon colonization of the region.

Environmental instability clearly imposed the most pervasive and enduring challenge to successful Mormon colonization of the Little Colorado River Basin. However, the expanding American frontier substantially compounded the problems encountered by these early pioneers (see Chapter 7). The arrival of the railroad in 1880 generated substantial non-Mormon immigration into the region which resulted in: (1) competing legal claims against Mormon-occupied land, (2) enhanced rangeland competition and deterioration, (3) increased lawlessness and violence, (4) open political conflict with non-Mormon interests, and (4) the arrest and imprisonment for polygamy of numerous local Mormon community leaders. Although the drain on local resources imposed by competing non-Mormon interests lasted only a few years, it proved critical and severely threatened the survival of several individual towns and, thus, the success of the colonization effort. However, tithing redistribution mitigated the negative consequences of adverse political conditions just as it had counteracted the drains imposed by natural environmental variability.

This book attempts to explain the success of Mormon colonization in the Little Colorado River Basin, and from this explanation broaden our understanding of successful Mormon colonization of the American West and of the evolution of human communities generally. Several researchers have already investigated Mormon settlement in the Little Colorado River Basin. Charles Peterson (1973) presents a narrative history of Mormon colonization along the Little Colorado from 1870 to 1900. Mark Leone (1979) utilizes historical developments among Little Colorado Mormon settlements to examine transformations in Mormonism as a theological system between the nineteenth and twentieth centuries. Kent Lightfoot (1980) examines sociopolitical developments among the early Little Colorado Mormon settlements to provide a model for understanding sociocultural development among prehistoric Pueblo societies in the Southwest. Directly or indirectly, each of these researchers has offered an explanation for successful Mormon colonization of the region. It might, therefore, reasonably be asked: if Mormon settlement in the Little Colorado River Basin has already been examined in detail by three professional researchers (see also Peterson 1970, 1976; Leone 1972, 1974) as well as by numerous local historians (cf. Tanner and Richards 1977; LeVine 1977; Westover and Richards 1964), what justification exists for undertaking yet another investigation of these two dozen seemingly inconsequential and economically marginal towns situated in a remote and isolated river basin in northeastern Arizona? The justification lies in the fact that no satisfactory scientific explanation of the settlement process yet exists, despite the existence of these studies.

A scientific explanation for the success of Mormon settlement in this region must account not only for the overall success of the colonization effort, but also for specific empirical developments associated with the settlement process, including: (1) the role of resource redistribution in successful colonization; (2) the differential success of Mormon efforts to establish a resource redistribution system capable of circumventing local environmental variability; (3) Local differences in community development; and (4) the impact of external factors on the settlement process. Furthermore, all of these developments should be consistently and parsimoniously explained as consequences of (that is, as testable predictions derived from) general theoretical principles (see Hempel 1965; Nagel 1979).

This has not been achieved, because previous researchers have either not adhered to basic scientific procedures in offering explanations for successful Mormon colonization of the region, or have applied scientific procedures erroneously and/or incompletely. They have not attempted to account for the general success of Mormon colonization in this arid and climatically variable region, together with specific developments attending the settlement process, within a precise, systematic and testable analytical framework derived from the application of general theoretical principles. To a large extent, previous investigators have failed to explain successful Mormon colonization of the region because they have adhered to traditional historical or anthropological approaches to the study of community development and social change which provide an inadequate basis for explaining specific empirical developments associated with the settlement process. The present analysis applies the analytical framework of general ecology. Ecological theory provides a more suitable analytical framework for explaining Mormon settlement in this region, because it systematically and parsimoniously accounts for successful colonization and the various empirical developments associated with the settlement process as consequences of general theoretical principles.

Previous Analyses of Little Colorado Mormon Settlements

Charles Peterson (1973) offers a detailed description of historical events surrounding Mormon colonization of the Little Colorado River Basin. Interrelating psychological, economic, social, political, religious and environmental considerations within an engaging historical narrative, Peterson provides a valuable interpretive account of the settlement process. However, while Peterson's account is both suggestive and informative, it cannot be accepted as an *explanation* of developments surrounding Mormon colonization of the region, at least not according to any scientific meaning of that term. Although Peterson provides numerous and valuable insights into the settlement process, his discussion remains a largely ideographic description of local events without broader systemic or theoretical implications. In addition, because his analysis is not clearly directed by an explicit consideration of general theoretical principles, his explanation of local developments remains both implicit and eclectic. Peterson repeatedly relies on descriptions of individual personalities and of religious commitment to both illuminate and explain historical developments. Personality characteristics of individual actors, environmental circumstances, Church resources, Mormon theology and other considerations are all blended indeterminately in Peterson's account to explain local differences in community development. To the extent that an ultimate causality is implied by Peterson, unique Mormon values and their individual and institutional manifestation are given primary consideration. Peterson's eclectic, implicit and non-systematic analysis results: (1) in a superficial and uneven consideration of important facts related to the settlement process; and (2) in nonparsimonious, logically unconnected and objectively unverifiable explanations of specific developments associated with Mormon settlement in the region. While Peterson's approach results in a valuable, insightful and highly readable account of the colonization effort, it leaves us with an explanation of this historical event which is unscientific (see Hempel 1965:231-244; Harris 1979:287-314).

Leone differs from Peterson in that he explicitly attempts to provide a systematic explanation for successful Mormon colonization of the region. In the course of their colonizing effort, Little Colorado Mormon pioneers suffered numerous dam failures (see Chapter 5). Adopting the analytical framework of cultural ecology (see Steward 1955; Bennett 1976; Netting 1977), Leone (1979:43-110) argues that successful Mormon colonization along the Little Colorado resulted in large part from the synergistic effect that dam failures and tithing redistribution had on regulating human environmental relations. Leone claims that while dam washouts represented substantial material losses for the communities involved, they also cleansed irrigation systems of silt accumulation and initiated a ritual cycle which resulted in: (1) the redistribution of tithing resources to affected individuals and towns; (2) intercommunity cooperation in dam reconstruction; and (3) an increased sense of unity throughout the regional Mormon population. According to Leone, the combined theological and material reinforcement of the ritual cycle sustained local commitment to the colonization effort in the face of overwhelming environmental obstacles and contributed ultimately to its success. The Mormon doctrine of being tried by God in the cause of building up his kingdom was the ideological rationale which completed the washout-rebuilding cycle (see Leone 1979:103-105). Leone argues that by initiating a ritual cycle which resulted in tithing redistribution and increased religious communion among numerous, widely-dispersed local settlements, recurring dam failures strengthened the sense of unity within the regional Mormon community and substantially advanced the success of the colonization effort.

Tithing redistribution was, indeed, critical to successful Mormon colonization of the region (see Chapter 6; Abruzzi 1989), and Leone has made a significant contribution to our understanding of the settlement process by linking the operation of this system to the successful colonization effort. However, his account of its adaptive role is flawed. It is neither systematic nor testable. His explanation does not involve an explicit application of general theoretical principles. Rather, it constitutes a functional explanation of human environmental relations modeled on Rappaport's (1968) analysis of highland New Guinea warfare (see Leone 1979:2-3). Following Rappaport's lead, Leone claims that tithing redistribution operated as part of an elaborate ritual cycle which functioned to maintain an equilibrium between the regional Mormon population and its environmental resources.

While there may be some validity to Leone's claim regarding the cleansing role of dam failures, his thesis regarding the functional role of dam washouts and tithing redistribution in enhancing Mormon unity and the success of the colonization effort is scientifically questionable. To begin with, it suffers from the same methodological problems associated with Rappaport's analysis of highland New Guinea warfare (see Friedman 1979) and inherent in functional explanations generally, ecological or otherwise (see Hempel 1959; Collins 1964; Vayda and McCay 1975; Bates and Lees 1979; Smith 1985). What needs to be shown--that the Mormon population was, in fact, in equilibrium with local environmental resources--is assumed rather than demonstrated. His thesis is also completely untestable. The purported integrative advantage gained from successive dam failures is non-quantifiable and, therefore, cannot be measured against the substantial material losses that such calamities imposed. In addition, Leone's thesis is contradicted by the facts. St. Joseph and Woodruff, the towns which suffered the greatest number of dam failures (see Chapter 5), were among the least developed communities in the region. Also, the lower valley, the subregion with the greatest incidence of dam failures, experienced the largest number of extinct settlements.

A major problem with Leone's functional analysis of Mormon colonization along the Little Colorado is that it occurs within a historical vacuum. He presents an analysis of Mormon settlement divorced from its empirical context. Individual historical events which significantly effected the settlement process, such as specific floods, dam failures and droughts, as well as the political conflict with non-Mormons, the advent of the railroad, and the arrival and land use policies of the Aztec Land and Cattle Company, are either ignored or discussed vaguely within a highly abstract synchronic and ahistorical analytical framework.

Consequently, Leone's analysis rests mostly upon generalizations about the region and about the regional Mormon population, both of which are described largely as undifferentiated entities. Little attempt is made to address subregional environmental differences and their precise relation to local variation in agricultural productivity, population size and other indices of community development. Leone also describes tithing redistribution as if it were an unchanging system which existed throughout the settlement period rather than as a system of resource redistribution which emerged following the failure of the earlier conjoint enterprises, and which itself evolved through time (see Chapter 6).

Because Leone treats the basin as a largely undifferentiated whole, he is unable to even address, let alone explain, the substantial local developmental differences that existed during the nineteenth century. Although at one point, Leone (1979:52) states that the basin can be divided into eleven environmental zones, he neither delineates these zones nor describes their distinctive physical characteristics. Moreover, he erroneously claims that local habitats were all equally suited to agricultural production (*ibid.*; see Chapter 4). Consequently, despite Leone's claim that he is offering an ecological explanation of the settlement process, his failure to provide detailed environmental data precludes any meaningful consideration of the purported ecological implications of his thesis regarding the regulative effect that dam failures and tithing redistribution had on human environmental relations.

An objective consideration of the scientific merits of Leone's thesis is further undermined by his reliance on such empirically undefinable concepts as "sanctity", "mental states" and "collective memories" as causal agents underlying the success of Mormon colonization (Leone 1979:*passim*). Leone's focus on "Mormonism", "Mormondom" and "Mormon Culture" as his central investigative units rather than on the specific local populations and settlements in the Little Colorado River Basin also represents a fundamental methodological weakness in his analysis of local developmental processes and underscores the essentially non-empirical focus of his investigation.[1] Where he does present local data, like Peterson, his discussion of critical information is largely eclectic, anecdotal and nonsystematic.[2]

Despite Leone's significant contribution to our understanding of the relationship between tithing redistribution and successful Mormon colonization of the region, his account of its specific role remains scientifically inadequate. Due to its extreme generality an its lack of a clear empirical focus, Leone's analysis of Mormon settlement in the region is unable to account for local differences in community development or for the differential success of early Mormon efforts to achieve a system of resource redistribution capable of counteracting the effect of local environmental variability. In the end, it explains very little regarding Mormon settlement in the region.

Kent Lightfoot (1980) also attributes successful Mormon colonization of the Little Colorado River Basin to the development of an effective system of resource redistribution. However, Lightfoot examines Mormon resource redistribution more clearly within its historical context. He attributes the success of tithing redistribution largely to the development of a two-tiered church hierarchy consisting of wards and stakes which, he argues, facilitated a more effective redistribution of surplus resources among the numerous widely scattered farming settlements in the region. Lightfoot also recognizes that the success of tithing redistribution resulted from local physical environmental differences. Moreover, he provides at least a partial account of what those differences were and how environmental diversity, together with tithing redistribution, contributed to successful Mormon colonization of the region. By placing tithing redistribution more clearly within its appropriate empirical context, Lightfoot offers an important revision of Leone's original insight. Yet, Lightfoot also fails to provide a systematic explanation for the evolution and success of tithing redistribution. First of all, Lightfoot contrasts the hierarchical church organization of stakes and wards associated with the regional system of tithing redistribution with the single-level organization of the early United Order settlements located in the lower valley of the Little Colorado River. In the latter system, he claims, individual towns were dependent largely on themselves.[3] By characterizing the lower valley settlements as separate and independent communal villages, he fails to discuss: (1) the development and importance of the conjoint enterprises in the redistribution of resources among lower valley settlements; (2) the fact that the lower valley settlements were organized as early as 1878 into the Little Colorado Stake; and (3) the fact that the United Order organizations did not prevail for long among the lower valley settlements. Lightfoot also does not account for the success of tithing redistribution in terms of the theoretical implications of the organization of this system in relation to local and regional environmental conditions. In the absence of general theory, he provides only a partial *substantivist* explanation for the success of Mormon colonization of the basin.[4]

The organization of tithing redistribution cannot itself be used as the explanatory agent for successful Mormon colonization of the region. Because that organization was part of the process of successful multi-habitat resource redistribution, it also needs to be explained. Lightfoot attributes the success of the two-tiered redistribution system as resulting largely from the local application of more encompassing Church institutions. While this is partially correct, he does not explain why this was the case in terms of general theoretical considerations. His thesis amounts, therefore, to a *diffusionist* or *historical particularist* explanation which fails to deal systematically with local causality (see Steward 1955:181-182; Harris 1968:250-318). Both the conjoint enterprises and the system of tithing redistribution were based on prevailing Church institutions. What needs to be explained is why the application of Church institutions proved successful in the one case and not in the other. This can only be achieved by using a systematic model which accounts for the differential success of the two systems in terms of general theoretical principles applied to local empirical conditions. More specifically, the general model needs to account for the differential success of the two systems of resource redistribution in terms of their distinct organizational characteristics in relation to the specific material environmental conditions each had evolved to counteract.

In summation, then, we are left without a satisfactory explanation of Mormon colonization of the Little Colorado River Basin, despite the fact that several studies of the settlement process have been undertaken and that issues directly related to the success of the colonization effort have been addressed. No satisfactory explanation exists because previous researchers have not adhered to the fundamental scientific procedures upon which a satisfactory explanation depends. At the very least, each has failed to explain empirical developments attending the settlement process as consequences of general theoretical principles.

The failure of Peterson's explanation derives largely from his eclectic research strategy and its failure to provide a coherent and systematic understanding of local differences in community development and of developmental processes operating during the settlement period. The failure of Leone's analysis lies: (1) in its non-empirical focus, (2) his use of a non-operational conceptual framework, (3) his failure to deal systematically with local environmental and developmental differences and the impact of external influences, and (4) his functionalist, ahistorical and synchronic analysis of the settlement process, particularly as it relates to the relationship between dam failures, tithing redistribution and successful colonization. The principal limitation in Lightfoot's analysis of Mormon colonization is that he fails: (1) to discuss the conjoint enterprises, and (2) to describe accurately and precisely the organization of the lower valley settlements. Furthermore, he does not explain the success of tithing redistribution in terms of conditions specified by general theory. Lightfoot also does not address local differences in community development or the impact that external factors had on the settlement process.

Explaining Community Development

Although Peterson and Leone employ two distinct analytical approaches to the study of Mormon colonization in the Little Colorado River Basin--one historical and the other anthropological--they draw similar conclusions regarding the success of the colonization effort. For both, successful colonization of this arid and climatically unstable region resulted ultimately from the superiority of cooperative Mormon values in overcoming obstacles imposed by a demanding environment. In this regard, Peterson's and Leone's explanations parallel other attempts to account for successful Mormon colonization. Indeed, a reference to distinct values has been the most pervasive and enduring explanation used by historians and social scientists alike to explain the general success of Mormon settlement in the American West (cf. McClintock 1921; Stegner 1942, 1964; Arrington 1958; 1964; Vogt and O'Dea 1953; O'Dea 1957; Meinig 1965; Vogt and Albert 1970; Arrington, Fox and May 1976).

However, a reference to unique values and institutions cannot provide a viable explanation of successful Mormon colonization in the Little Colorado River Basin or elsewhere. To begin with, such explanations ignore the material context which select for the origin and persistence of distinct values within a group (cf. Barth 1969; Nelson 1973; Gelfand and Lee 1973; Abruzzi 1982). A reference to unique values also yields a distinct explanation for each social group and, thus, runs counter to the scientific goal of developing increasingly general and, thus, more powerful and parsimonious explanations. Such explanations also deal only with proximate rather than ultimate causes for the success of the colonization effort (see Barash 1977:37-39; Abruzzi 1988). Finally, a reference to cooperative Mormon values cannot explain the considerable developmental variation that existed among culturally homogeneous Mormon settlements. Nor can it account for the greater success of tithing redistribution as a mechanism of environmental regulation contributing to the Little Colorado colonization effort compared to the conjoint enterprises, especially since the former, unlike the latter, was fundamentally based on institutional arrangements which facilitated individual financial gain (see Chapter 6) while the latter was based on communal values and institutions. A scientifically meaningful explanation of Mormon colonization must account for the substantial developmental variation that occurred among Mormon settlements, including both the developmental differences which existed among individual towns and the differential success of attempts to establish a viable multi-habitat system of resource redistribution. Moreover, the explanation must account for this variation consistently and parsimoniously as a consequence of general theoretical principles.

As previously indicated, the inability of Peterson's interpretive account to explain Mormon colonization in the Little Colorado River Basin derives largely from his non-systematic analysis of the data and from his ultimate reliance on the facts of the situation (with only implied or plausible interconnections) to explain themselves. This same analytical approach has characterized most historical analyses of Mormon colonization in the American West (cf. Stegner 1942, 1964; Arrington 1958) and has typified historical explanation generally. Indeed, simply conveying plausibility and rationality has been proposed by several scholars as the only legitimate goal of historical explanation (cf. Dray 1974; Collingswood 1974). May Brodbeck (1962:254), on the other hand, has argued that "there is no such thing as 'historical' explanation, only the explanation of historical events." From this perspective, the form of historical explanation should not differ in kind from that used to explain empirical developments in other fields (see Hempel 1942). Historical "events" should be explained through the systematic application of general theoretical principles which account for local empirical developments within a parsimonious, predictive and testable analytical framework (see Popper 1957; Hempel 1965; Nagel 1974, 1979). This does not mean, however, that a specific historical event can be explained in its entirety. Historical developments are no more capable of complete explanation than are events in the physical or biological realm or, for that matter, in any area of empirical research (see Popper 1957:140).

Anthropologists and other social scientists have also been interested in explaining social change, including the evolution of complex human communities (see Harris 1968; Nisbet 1969). However, despite more than a century of investigation and theory development, social scientists have yet to develop a testable model of social evolution and community development which can systematically explain successful Mormon colonization of the Little Colorado River Basin and the various empirical developments associated with the settlement process. The inability of social evolutionary theory to explain this historical event derives largely from four methodological problems.

First, explanations of social evolution and community development have been largely typological, proposing empirical generalizations regarding the sequence of developmental stages abstracted from the ethnographic record (cf. Tylor 1871; Morgan 1877; White 1959; Sahlins and Service 1960; Service 1971; Flannery 1972; Faris 1975; Rose 1981; Kottak 1982). Such explanations lack the essential scientific concern for applying a systematic general theory to make testable predictions about specific local empirical phenomena. Too often, theoretical relationships are either ambiguous (cf. Carneiro 1962, 1967, 1968; Adams 1966; Flannery 1972) or derive exclusively from a local ethnographic context rather than from a consideration of that context in general theoretical terms (cf. Foster 1967; Bennett 1969; Vogt and Albert 1970).

Second, anthropological explanations of social evolution have largely utilized synchronic or cross-sectional data (cf. Sahlins 1958; Carneiro 1962, 1967, 1968; Kottak 1982). Inferring diachronic processes from such data is methodologically questionable because empirical developmental relationships can only be established through the use of time-structured information (see Barth 1967; Graves, Graves and Kobrin 1969; Plog 1973).

Third, general theories of social evolution and community development largely focus on *cultures* and *societies* (cf. White 1959; Sahlins and Service 1960; Flannery 1972; Kottak 1982), neither of which constitute viable analytical units for investigating local community development. Besides being non-operational, cultures and societies are inappropriate units for analyzing evolutionary/developmental processes (Vayda and Rappaport 1968). The selective forces which generate community development operate on specific local populations adapting to surrounding regional systems (see Ricklefs 1987). Consequently, it is upon local populations rather than upon cultures and societies that the analysis of social evolution and community development must concentrate.

Fourth, general anthropological explanations of social evolution have largely viewed human communities as closed systems which develop independent of local environmental considerations (cf. White 1959; Service 1971; Bennett 1976). While such explanations explore the internal dynamics of changing social systems, they largely ignore the developmental role of external factors in community development (see Lees and Bates 1984).

For these reasons, anthropological explanations of social evolution and community development cannot explain the differential development of nineteenth century Mormon settlements in the Little Colorado River Basin. Empirical generalizations regarding the transition from band to tribal or from traditional to modern societies are not applicable to these communities. Likewise, explanations of community development based on unique cultural characteristics cannot explain the considerable developmental variation that occurred among these culturally homogeneous settlements. A scientifically meaningful explanation of community development in the basin can only be provided through the systematic application of a general model which predicts local developmental variation as a consequence of the theoretical relationships linking empirically definable social and environmental variables. Several anthropologists have focused on local populations adapting to specific external environments (cf. Steward 1955, Sahlins 1958; Netting 1968; Bennett 1969; Meggers 1971; Kottak 1982). However, in such cases environments have generally been viewed typologically, that is as "things" (Athens 1977), rather than as complex, dynamic and variable systems. Consequently, general models have not emerged from such studies which systematically interrelate environmental and social variables within a predictive and testable theoretical framework.

Ecology and Community Development

The development of complex human communities is an evolutionary process which necessarily occurs through time and which is substantially constrained or advanced by local and regional conditions of resource availability. An explanation of Mormon colonization in the Little Colorado River Basin requires, therefore, that local differences in community development be explained by a general theory which systematically accounts for variation in community development as a consequence of distinct population/resource relationships and within an explicit, diachronic analytical framework. The general theory employed must also consistently and parsimoniously explain the overall success of Mormon colonization, as well as specific empirical developments attending the settlement process, including: (1) local differences in community development, (2) the role of tithing redistribution in successful Mormon colonization, (3) the differential success of Mormon attempts to develop an effective multi-habitat resource-flow system, and (4) the impact of external factors on the settlement process.

In order to better explain successful Mormon colonization of the Little Colorado River Basin, the following analysis employs an ecological model and a diachronic research strategy to account for specific empirical developments attending the settlement process. The model employed is an adaptation of the general model developed by plant and animal ecologists to explain the evolution of complex ecological communities.

Considerable controversy surrounds the application of ecological concepts to human populations (see Bennett 1976; Moran 1984; Abruzzi 1987). While numerous researchers in several disciplines have utilized ecological concepts and principles to explain human social behavior (cf. Barth 1956; Rappaport 1968; Gall and Saxe 1977; Leone 1979; Winterhalder and Smith 1981; Abruzzi 1982, 1987, 1989), others have rejected such applications as largely naive and inappropriate uses of biological concepts (cf. Vayda and McCay 1975; Bennett 1976; Lees and Bates 1984; Smith 1984; Young and Broussard 1986). Ecological concepts have indeed been misapplied in human ecology. However, their misapplication has occurred not because such concepts are inherently inapplicable to the human species, but rather because they have largely been applied incorrectly. In most cases, ecological concepts have been extended to human populations wholly disconnected from encompassing theoretical systems. They have largely served as metaphors underlying a functionalist view of human-environmental relations (see Vayda and McCay 1975; Smith 1984). Moreover, the term "ecology" has generally been applied in a restricted sense in social analysis, even by anthropological ecologists, to refer simply to relationship which exists between a human

Introduction 13

population and its natural environment. Ecology has generally not served as a body of theory leading to testable hypotheses regarding the organization and evolution of human communities. The ultimate goal of the present investigation is to go beyond a mere metaphorical and environmentalist application of general ecology to Mormon colonization of the Little Colorado River Basin to show that general ecology provides a meaningful and productive theoretical framework for explaining not only Mormon colonization in this region and elsewhere but also, potentially, the evolution of complex human communities.

The application of ecological theory to the Little Colorado Mormon settlements rests on the assumption that human communities are ecological communities through which energy flows and by which population/resource relationships are regulated (see H. Odum 1971; Little and Morren 1976; Abruzzi 1982:13-14; Chapter 3).[5] Any system which contains living organisms constitutes and ecological system (see Margalef 1968; H. Odum 1971). Consequently, human communities should be considered ecological systems. Several important similarities exist between human and nonhuman communities which indicate the value of their being analyzed from a unified theoretical perspective: (1) both constitute material systems through which energy flows and by which population/resource relationships are mutually regulated (cf. Margalef 1968; E. Odum 1971; H. Odum 1971; Little and Morren 1976); (2) both contain a potentially high degree of functional diversity which is ultimately dependent on continuous inputs of energy from external sources (H. Odum 1971); (3) both ultimately transform the potential energy available in their environments into social organization; (4) both contain organizational units which vary in size and composition as a result of spatio-temporal changes in the abundance and distribution of resources (see Wilson 1968; Kummer 1971; Abruzzi 1982) and (5) the processes which underlie the division of labor (i.e., resource partitioning) are central to the evolution of both types of communities (cf. Harris 1964; Blau 1967; Levins 1968).

Viewed from this perspective, an industrial city is just as much an ecological system as is a tropical rain forest. Regulated by energy flows which determine population distribution and functional specificity, the settlement pattern and community organization that evolve in industrial-urban areas are distinct from those present in human communities which subsist primarily by farming or by exploiting undomesticated plant and animal resources. Due to the greater efficiency produced by the centralization of productive and distributive functions in industrial systems and by the clustering of consumer functions in ecological systems generally, the population concentration and consequent complex organization associated with such urban areas constitute predictable responses to the maximization of resource exploitation through industrial production.

As previously indicated, some researchers have rejected the extension of general ecological theory and concepts to human populations on the grounds that ecology is a biological science, whereas human populations adapt through socioculturally acquired behaviors. However, it must be noted that ecological principles are independent of a particular community's specific biological composition, because they are a consequence of the energetic, not biological, relationships which prevail within and between systems subject to selective pressures. Indeed, ecological systems have been most effectively modeled as energy-flow systems whose organization and evolution are determined in large part by thermodynamic principles (cf. Margalef 1968; H. Odum 1971; E. Odum 1971:37-85; Little and Morren 1976). Furthermore, while the properties (i.e., the structure) of any specific ecological community are to a large extent a function of the organisms which comprise it, the principles (i.e., the processes) which determine the organization and evolution of ecological communities apply to all communities, regardless of their specific biological composition. If ecological processes are independent of the specific biological composition of a community, so also must be the principles and concepts used to explain the organization and evolution of such communities. They should apply equally to single and multispecies communities, to terrestrial and aquatic communities, as well as to human and nonhuman communities.

It may be scientifically (i.e., nomothetically) preferable to approach human social systems as a subset of more general ecological systems, subject to the same general theoretical principles, than to continue to regard human communities as analytically distinct from all other social systems. From the perspective of theory development, it matters less whether human and nonhuman communities are substantively distinct, than whether general ecological concepts and principles account for comparable empirical developments in both types of systems. A single explanation for the organization and evolution of human and nonhuman communities is more parsimonious and more powerful than, and thus preferable to, two distinct explanations: one for human communities and one for the remainder of the organic world.

However, the application of general ecological concepts and principles to human populations must go beyond the simple relabeling of social phenomena with ecological terms. It must be based on a recognition that similar *processes* operate in physically distinct and unrelated systems (see Ashby 1956; von Bertalanffy 1968; Day and Grove 1975; Rapport and Turner 1977; Alexander and Borgia 1978; Abruzzi 1982). The approach is, therefore, *systemic* not reductionist. The goal must be to determine if the same concepts and principles used to explain the evolution of complex nonhuman communities can be applied to account for empirical developments associated with the evolution of human communities as well. However, to achieve this goal, ecological concepts and principles must be applied within a predictive and testable theoretical framework. It is only when the processual models of general ecology are applied formally and explicitly to human communities that testable predictions can be generated and that the applicability of these models can be evaluated objectively. Consequently, in order to explain and not merely describe or heuristically illustrate patterns of community development, ecological theory must account for specific empirical developments as a consequence of predictions derived from general theoretical considerations. However, as in any experimental situation, general theoretical concepts and

Introduction 15

principles must be modified and operationalized to fit the specific empirical problem and context under investigation.

Ecological communities evolve primarily as a response to population growth and to changing conditions of resource availability (see Brookhaven National Laboratory 1969; Whittaker 1975; Cody and Diamond 1975). The model employed here deals explicitly with the relationship between population growth, community productivity and functional community diversity and systematically connects changes in these three community parameters with variations in resource availability. The principles which underlie this model apply to all ecological communities, regardless of their specific biological composition. Consequently, while it is sufficiently operational to be applied to a specific empirical community, the model is also general enough to be exported to a variety of ecological communities, including early Mormon settlements in the Little Colorado River Basin.

In order to explain Mormon colonization of the Little Colorado River Basin, the following investigation begins with an overview of Mormon settlement in the region. Chapter 2 provides both a summary of Little Colorado Mormon settlement history and a detailed description of the variation in community development that resulted. Concentrating on the adaptive problems encountered by early Mormon pioneers, Chapter 2 illustrates the extent of variation in population size, productivity, stability and functional diversity achieved by individual towns.

Once an overview of the settlement process has been presented, attention will turn to the ecological model employed to explain developments associated with the settlement process. Chapter 3 describes the succession model, its implication for human communities generally, and its application to the specific settlements under investigation. The chapter begins with a discussion of the niche concept and the effect that different environmental conditions have upon niche specialization and differentiation. Approaching ecological communities as energy-flow systems, Chapter 3 stresses the impact that external factors have on the cost of community maintenance and the developmental role that such contingencies play as either subsidies to or drains upon community development. The chapter interrelates the role of environmental productivity and stability in determining functional community diversity and the regulative capacity of complex ecological communities within a synthetic theoretical model.

Having presented an overview of the settlement history and a description of the general model to be used to explain developments associated with the colonization effort, the focus turns to specific aspects of the settlement process which bear directly upon conditions specified in the ecological model. Chapter 4 presents a detailed description of the natural environment of the Little Colorado River Basin. Spatial differences in topography, precipitation, temperature, soil quality, plant communities, and surface water quality and availability are discussed, together with their implication for human resource exploitation. Sub-regions are then distinguished within the basin where similar environmental conditions produced comparable developmental histories among individual settlements.

Irrigation has been a fundamental prerequisite for successful farming in this region due to the arid and variable climate which characterizes the Little Colorado River Basin. Variation in the availability of irrigation water was the primary factor determining fluctuations in agricultural productivity and population size among early Mormon settlements in the basin, as well as the frequency of dam failures. In addition, irrigation systems constituted the principal infrastructural investment in each town throughout the nineteenth century. Consequently, Chapter 5 examines the history of dam constructions and failures for individual settlements in order to determine both the abundance and stability of local water supplies and the maintenance costs endured by individual towns. The number of dam failures for individual settlements between 1876-1900 ranged from zero to 13, with a clear relationship existing between the incidence of dam failures and an individual town's sub-regional location.

The Little Colorado River Basin contains numerous, widely separated habitats possessing unique patterns of resource availability. In order to circumvent the resource instability associate with individual habitats, early Mormon pioneers attempted to integrate the productivity of several habitats into a single resource--flow system. As indicated previously, their efforts were not equally successful. Chapter 6 examines ecological and other material conditions underlying the differential success of Mormon efforts to establish a viable multihabitat resource-flow system integrating the economies of the various settlements in the basin. This chapter demonstrates that certain Church institutions provided a pre-adapted organizational apparatus facilitating the formation of a regional resource redistribution system. By materially subsidizing the evolution of a regional resource-flow system, such institutions contributed prominently to the success of the colonization effort.

As already indicated, the advent of the railroad stimulated regional economic development and caused the immigration of numerous non-Mormons into the area. Expropriation of resources, competing land claims, political persecution, thievery and violence were the principal stresses endured by early Mormon settlers subsequent to the railroad's penetration of the region. Chapter 7 focuses on the regional Mormon population's principal competitors and assesses their impact on the settlement process. This chapter reiterates the importance of Church institutions in subsidizing the settlement process, this time by offsetting the drains imposed upon community development by competing non-Mormon interests rather than by an unstable environment.

The book concludes with an explicit application of the general ecological model presented in Chapter 3 to account for specific empirical developments associated with Mormon colonization of the region. In Chapter 8, differences in community development among individual Mormon settlements are tested against predictions derived from ecological theory. In addition, sub-regional differences in community development, the role of tithing redistribution in successful colonization, the variable success of Mormon efforts to establish a multihabitat resource-flow system, and the impact of external factors upon the settlement process are explicitly evaluated in terms of predictions derived from the general ecological model. All of these empirical developments are consistently and parsimoniously explained by this single model.

It will be shown that general ecology provides what has previously been lacking--a systematic and parsimonious explanation for the success of Mormon colonization of the Little Colorado River Basin and for specific empirical developments associated with the settlement process. The monograph concludes by discussing the advantages gained from the successful application of ecological theory to Mormon colonization in the Little Colorado River Basin. These include: (1) ecological theory's ability to systematically and parsimoniously account for several specific features of the settlement process left unexplained by previous researchers and unexplainable by traditional historical and anthropological explanations; (2) the exportability of the ecological model to other ethnographic situations; (3) the implications that successful application of the model in this region has for our understanding of the ultimate role of material and ecological considerations in successful Mormon colonization of the American West; and (4) the broader theoretical advantages to be gained by analyzing human communities as empirical variants of more general ecological systems.

NOTES

1. Leone's nonempirical focus is further illustrated by the fact that barely one-third of his book specifically discusses the Little Colorado settlements. He is, in fact, more concerned with explaining changes in Mormon theology than with accounting for empirical developments along the Little Colorado. The Little Colorado settlements represent a metaphor for his analysis of Mormon colonization generally. However, metaphors cannot provide a systematic and predictive explanation of social behavior.
2. Leone (1979:229-241) provides considerable detailed tithing data in an extensive appendix to his book. However, this information is not systematically incorporated into his analysis of the settlement process.
3. The Mormon Church is divided administratively into stakes and wards, which may be compared to diocese and parishes respectively in the Roman Catholic Church. During the nineteenth century, each Little Colorado settlement contained one ward. Wards in the lower valley of the Little Colorado River were initially organized into the Little Colorado Stake, with the remaining wards included within the Eastern Arizona Stake. In 1887, local wards were reorganized into the Snowflake and St. Johns Stakes, containing the western and eastern settlements respectively.
4. The term *substantivist* is used here as applied in economic anthropology in contrast to a formalist analytical approach (see Schneider 1974).
5. The term *regulation* has a problematic history in ecology, because it has generally been used to imply the internalized maintenance of a balance between populations and resources in ecological systems (cf. Margalef 1968; Rappaport 1968; Odum 1969; Leone 1979; Patten and Odum 1981), a claim which has been shown to be both incorrect and

undemonstrable (see Engelberg and Boyarsky 1979; Moran 1984). The term is not used here to imply the existence of a homeostatic equilibrium in ecological systems, but rather merely to the existence of processes through which the actions of populations positively influence the availability of resources within a community (see Chapter 3, footnote 14).

CHAPTER 2

Colonizing the Little Colorado River Basin

In 1876, under the direction of their church in Salt Lake City, approximately 500 Mormon pioneers founded four settlements along the lower valley of the Little Colorado River. Serving as footholds within the region, these initial settlements provided a foundation from which numerous additional colonies were established throughout the remaining river basin. While not comprising the first settlements within this region,[1] the Mormon villages established during the nineteenth century represented the effective onset of Anglo colonization of the Little Colorado River Basin. Most of the towns currently located within this basin were established as part of the nineteenth century Mormon expansion.[2]

The present chapter provides an overview of the settlement process. Focusing on the difficulties encountered by early Mormon pioneers in their attempt to establish viable agricultural communities within this variable natural environment, the chapter concentrates on the temporal variation that occurred in population and agricultural productivity among the various settlements between 1876 and the close of the nineteenth century. The individual settlements under consideration differed markedly among themselves regarding the size and variability of these two critical community components. Significant differences also existed between towns with respect to the functional complexity that each had achieved. More detailed examination of special features of the settlement process and of their impact on community development within the basin follows in subsequent chapters.

INITIAL COLONIZING EFFORTS

The first Mormon pioneers to settle the Little Colorado River Basin entered the area under the "call" of their church leaders in Salt Lake City.[3] Selling whatever property they owned, these original immigrants came from various parts of Utah and ventured south in organized companies into an unknown territory in an effort to expand the southern frontiers of Zion.[4] Mormon colonization along the Little Colorado River was part of the conscious expansion of the Mormon domain; through prior occupation of every arable valley, church leaders hoped to block Gentile (i.e., non-Mormon) settlement throughout the Mountain West, from Canada to northern Mexico (Arrington 1958:84-95). The vitality of this Mormon expansion was impressive; within ten years after the founding of Salt Lake City in 1847, ninety-six additional Mormon settlements had been established (*ibid.*:88), and by 1900 this number had been extended to over 500 (*ibid.*). Joseph City, the only original Mormon colony in the Little Colorado River Basin to survive, was the 318th colony in this expansionary process (Tanner and Richards 1977:10). Successful colonization of the Little Colorado did not come easily, however; in addition to the fact that only one of the four initial colonies survived, two previous, unsuccessful attempts to settle the lower valley have been recorded. In both cases, one Mormon and the other non-Mormon, the colonization effort failed despite the fact that the pioneers were organized into colonizing companies in order to maximize their chances for success.

Several scouting expeditions were sent into the Little Colorado River Basin by church leaders in the early 1870's, mostly under the direction of Jacob Hamblin, the famous Mormon scout and missionary to the Navajo and Hopi. As a result largely of Hamblin's reports, "a substantial company, well-equipped, containing many seasoned pioneers" (Tanner and Richards 1977:12) was organized which arrived at the Little Colorado River (near the Black Falls) towards the end of May in 1873. This company contained 109 men, 6 women, 1 child, 54 wagons with supplies and 112 animals (unspecified) (Porter n.d.d.; McClintock 1921:135). From this point, an exploring party was dispatched 136 miles upstream (*ibid.*) to the approximate vicinity of present-day Joseph City and Holbrook. Horton Haight, the captain of the company, characterized the Little Colorado as "a small amount of salty mineral water and quicksand and mineral bottom" (quoted in Tanner and Richards 1977:12). Describing the conditions which they encountered at the furthest point of their excursion, Haight reported:

> We see nothing better ahead of us. The river closes in again above the upper bottom, the hills are red and bare, we have had heavy winds nearly every day, at times enveloping us in Storms of Sand (sic.). With the poor feed and bad water, our animals are failing...passing, we noticed the water failing; at the upper falls there was but little. At the lower or black falls (sic.) it had stopped running (*ibid.*).

This company returned to Utah, even though they received a letter from Brigham Young directing them to proceed with their colonization along the Little Colorado River.

In 1876, the American Colonization Company of Boston organized two companies (about 100 men) to settle in the lower valley of the Little Colorado River (McClintock 1921:149; Peterson 1973:1-2). These settlers were inadequately prepared, however, having been given glowing and inaccurate descriptions of the Little Colorado River Valley. Finding several Mormon settlements already established in the area,[5] these two companies continued westward founding Flagstaff, Arizona instead.

Independent of these two organized efforts to colonize the Little Colorado River Basin, there existed an unorganized expansion from New Mexico into the eastern margin of this territory. During the early 1870's, New Mexican ranchers expanded into the pastures and valleys of the Little Colorado River Basin.[6] This Anglo colonization occurred in conjunction with a vigorous though undirected expansion of Hispanic populations (mostly sheepherders) up and beyond the broad river valleys extending from the Rio Grande Valley in central New Mexico (Meinig 1971:27-32). The Hispanic expansion led to the formation of two settlements--Concho and St. Johns--which pre-date all Anglo towns in the basin, and which have retained a significant Spanish-speaking population to this day. New Mexican businessmen also participated in the expansion which took place along the eastern portion of this territory, both at St. Johns and further south in Round Valley where a growing Anglo population had led to the formation of Springerville. While this gradual occupation by disparate and independent efforts had its impact upon the development of the basin, the effective colonization of this region was largely a function of Mormon expansion. By the early twentieth century, when most of the extant communities had been founded, over two dozen settlements existed within the basin. The majority of these were Mormon towns.

ESTABLISHING SETTLEMENTS: 1876-1880

Although extensive Mormon immigration into the Little Colorado River Basin continued throughout the latter part of the nineteenth century, not all of this period displayed the same intensity of community formation. Most Mormon towns in the basin were founded between 1876 and 1880. Subsequent years witnessed the settlement of a few smaller outlying towns as well as the immigration of additional Saints[7] to bolster already established locations. Owing to the difficult nature of settlement along the Little Colorado, extensive emigration occurred, and substantial reinforcement was needed to maintain the viability of certain towns.

Peterson (1973:16-17) divides the principal settlement period into three distinct geographical and chronological phases. The initial phase was represented by the establishment of four settlements--Obed, Sunset, Brigham City and St. Joseph--along the lower valley of the Little Colorado in 1876 (see Map 2.1).[8] A second series of Mormon colonies--including Woodruff, Snowflake, Taylor, and Showlow--were founded along Silver Creek in 1877 and 1878. The settlement period ended with the establishment during 1879 and 1880 of Mormon populations at St. Johns, Springerville, Eagar, Alpine and other locations along the upper reaches of the Little Colorado River. Each phase of the settlement process egressed from the one which preceded it. The finances, supplies, personnel, organization and emotional support which made each subsequent settlement possible came, in large part, from the surpluses provided (often at great expense) by the previously settled communities. Although the interdependency which accompanied the settlement process made the establishment of numerous communities possible, overextension of the settlement process within a highly

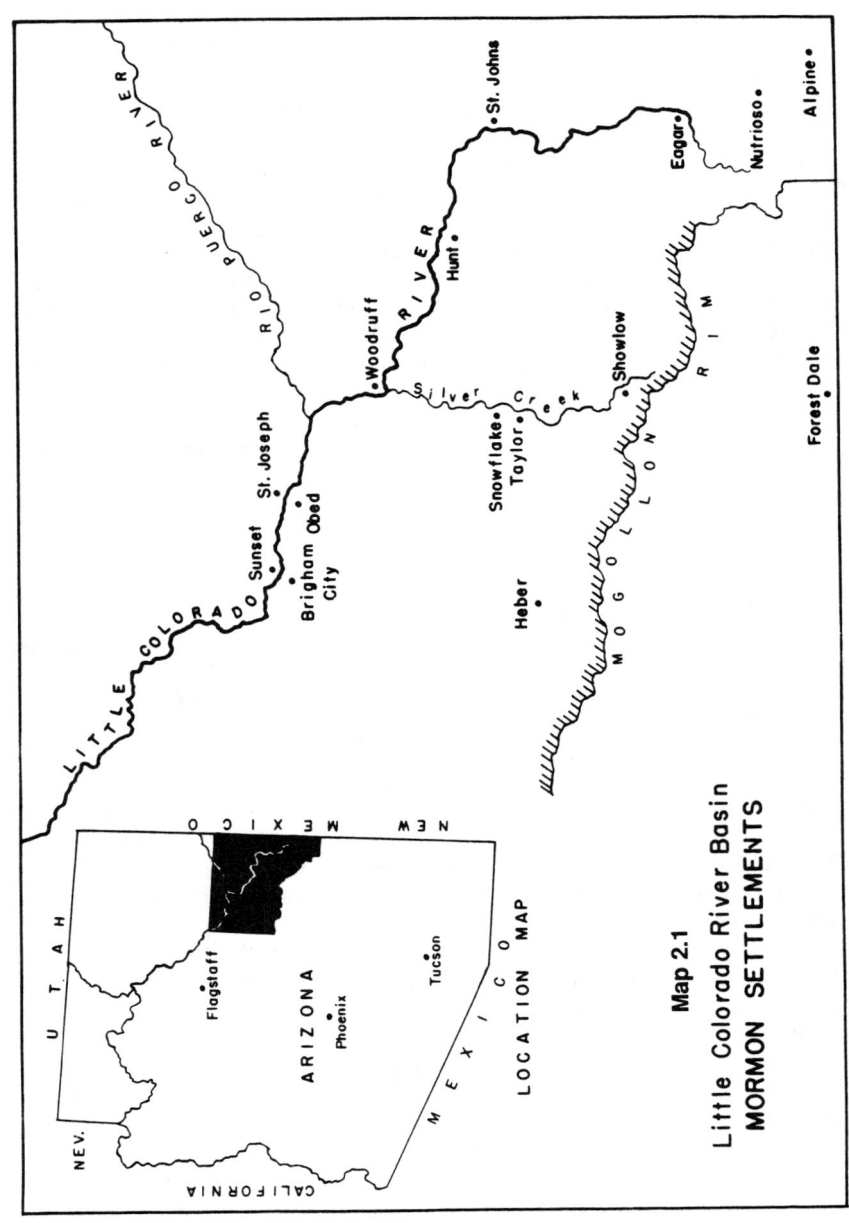

Map 2.1
Little Colorado River Basin
MORMON SETTLEMENTS

variable climatic regime placed considerable stress on the resources of donor and recipient communities alike.

1876

Subsequent to the failure of the Haight expedition, a second Mormon colonizing effort was organized in 1876. Calls went out to 200 men who were then organized into four companies,[9] each comprising 50 men and their families (McClintock 1921:138).[10] The leading teams arrived at their destination on March 23, 1876, with others joining them during the successive weeks. Recognizing the urgency of producing a crop the first year--since food could otherwise only be obtained by hauling it with wagons over great distances from either Kanab, Utah or Albuquerque, New Mexico (Porter n.d.a.:60)--work commenced immediately on the arduous tasks of planting fields, building dams and constructing irrigation canals. At St. Joseph,[11] the first plowing began on March 25th, only two days after the initial teams arrived at this location (McClintock 1921:140). Within two days of the first plowing, surveying of the irrigation ditch had begun, and on the following day the first logs to be used in the diversion dam were cut. By April 3rd, the first wheat had been sown (*ibid.*).

Three dams were constructed: one each for Sunset and Brigham City, and a combined dam for Obed and St. Joseph. The latter dam has been described as follows:

> Driftwood and brush, battened down by hundreds of tons of rock from a nearby quarry, were to be placed on the sandy bottom of the river. The diversion dam was designed to raise the water twelve or fifteen feet to force it over the river bank into the irrigation ditches. Nearly one hundred men in the two camps set to work on the project (Tanner and Richards 1977:41).

With construction beginning, as indicated, on March 28, the St. Joseph-Obed dam and ditches consumed nearly two and one-half months to build, and water was turned into the fields for the first time--and with much celebration--on June 6 (*ibid.*). Some disagreement exists concerning the exact height of the dam, with claims of 9 feet (Peterson 1967:39) and 12 feet (Porter n.d.a.:60) being made. The other dimensions do not appear to be in dispute; the dam was 180 feet long and 60 feet thick (Peterson 1967:39). Rulon Porter[12] (n.d.a.:60) quotes an early St. Joseph chronicler who recorded the following account of the labor expended on the combined St. Joseph-Obed dam.

> Our company done (sic.) 561 days hand labor and 136 days team labor on the dam. The total hand and team labor on the dam by both companies...was 1696 days. We also done (sic.) 647 days hand and 218 days team work on the ditch.

At the wage rate established at the time of $2.00 per day for a man and $1.50 per day for a team, the cost of this initial dam was recorded as $6,244, "divided approximately equally between the two colonies (*ibid.*)."[13] While no clear indication to this effect exists in any of the sources, the final cost as quoted would have to have included the cost of the irrigation ditches for both settlements. The ditch providing water for St. Joseph was three miles long, while a six-mile ditch was needed to reach Obed (*ibid.*). Sunset's dam, by comparison, was reported to have been 200 yards long with a three mile ditch (Jensen n.d.a.:4/28/1876), although its other dimensions were apparently comparable to those of the St. Joseph-Obed dam. Such was the length required to divert the Little Colorado River's water from its sandy bed.

Each settlement managed to plant about 50 acres of wheat, plus some corn; other demands upon their labor, including the building of a fort[14] and the digging of wells (but most importantly the construction of dams and ditches), did not leave sufficient time to plant much else. However, as the rains were late in coming and the Little Colorado was experiencing its seasonal decline (see Chapter 4), insufficient water was available for irrigation purposes. A "superabundance" of dust was reported for this period (Jensen n.d.a.:8/31/1876), and Brigham City recorded an inability to get any water out of the Little Colorado River and up over the dam due to the insufficiency of water in the stream (Jensen n.d.a.:4/6/1876). As late as July, the crops at Sunset and Brigham City were reported to be still "in the ground on account of the ground being so dry" (Jensen n.d.a.:7/16/1876), and no sign of green vegetation existed near either camp. Fifty acres of wheat were reported to have perished at Sunset alone due to the extended dry spell, forcing settlers there to plant an additional 20 acres of wheat and 75 to 80 acres of corn (*ibid.*:8/31/1876).

The first rain recorded by the settlers finally fell on July 16, and on July 17 a torrential flood completely swept away the joint dam of St. Joseph and Obed and severely damaged the dams at Sunset and Brigham City (*ibid.*:7/17/1876). A second flood visited the settlements on August 25 and completely destroyed what was left standing by the first flood, as well as what had been constructed during the interim. This second flood devastated Brigham City's dam, cutting a new channel in the river over 150 feet from its original course. Sunset's dam was also destroyed, and fields everywhere were flooded.

Harvests throughout the four settlements were a failure. Only Sunset is generally acknowledged to have produced a crop, consisting principally of about 75 bushels of grain and a few melons (cf., Jensen n.d.a.:12/31/1876; Peterson 1973:19). Scattered reports suggest that St. Joseph may have managed to raise about 10 bushels of corn and 54 bushels of wheat, possibly though the construction of "sand dams" (Porter n.d.a.:60).[15] In any case, the harvest was completely inadequate to sustain the population of the four settlements until the following year. Climatic variability had taken its toll this very first year, as the settlements sustained crop losses from the extremes of both drought and flooding during the same agricultural season.

Owing to their meager supplies, the colonists voted to permit the majority of their members to leave for Utah, where they would remain for the winter. These emigrants would return to the Little Colorado in the spring with their families and additional supplies. A reduction in the size of the population along the Little Colorado reduced the drain on the limited resources available there. The return to Utah also permitted the acquisition of scarce provisions and the recruitment of additional labor to renew the colonizing effort the following season.

Nearly 75% of St. Joseph's population returned to Utah, leaving only 6 single men plus 7 married men with their families (Tanner and Richards 1977:36). Twenty-three men had left St. Joseph for Utah (McClintock 1921:140), and of the 73 persons who had settled St. Joseph that year, only 22 persons remained. Obed's population was reduced from 123 to 22 as well (Tanner and Richards 1977:134), including only 9 men (Warner 1968:8). Sunset's population was cut in half, comprising only 22 men, 15 women and 20 children. The aggregate population of these four settlements, thus, dwindled from nearly 500 persons the previous spring to slightly more than 100 persons the following winter. Most of those who left for Utah in the fall of 1876 never returned.

The 1876 harvest was still unable to sustain the remaining population, even though most of the settlers had abandoned the Little Colorado River settlements that winter. Several teams had to be dispatched to southern Utah, particularly to Kanab (about 400 miles), in order to offset food shortages and to obtain seed for the 1877 growing season. Provisions received from Kanab included 3,336 lbs. of flour, 505 lbs. of corn meal, 186 lbs. of shorts, 69.5 lbs of molasses and 177.75 lbs of pork (Joseph City United Order n.d.a.:3/19/1877).

While the inadequate crops produced from the first year's labor did not inhibit Mormon colonization of this region, it did provide these settlers with an experience which was to become all too familiar in the years ahead. Many months of hard labor would produce crops which were insufficient to sustain the colonists even to the following harvest, let alone provide the surpluses needed to build increasingly diverse and stable communities. The wagon road to Utah was to become a well-traveled highway, providing a continuous subsidy without which the early settlements along the Little Colorado could not have been maintained.

1877

The following year proved more successful for colonists along the Little Colorado. Not only were harvests more plentiful, but 1877 marked the initiation of settlement along Silver Creek and, thus, the expansion of Mormon colonization within the basin. These gains were not achieved without a struggle, however.

In 1877, Obed and St. Joseph constructed separate dams. The settlers at St. Joseph located an appropriate site two miles upstream from that of the dam destroyed the previous year. In addition to building a new dam and repairing 3 miles of existing ditch, the settlers at St. Joseph now had to construct an additional 2 miles of new ditch. At the prices of hand and team labor referred to earlier, this new diversion dam and the accompanying ditch work was reported to have cost $3,328.50 (Porter n.d.a.:355).[16] The entire labor was performed by 18 to 20 men (*ibid.*), and water was turned into the fields on May 18, 1877 (Porter n.d.b.:5).[17]

Not much information was recorded concerning the events of 1877. Although more productive than the previous year, enough data exists to suggest that this year was not as productive as it might have been. On August 22 of that year,

> A disastrous flood visited the lower settlements,[18] carrying off large quantities of cut grain shocked in the field and badly damaged what was stacked. Brigham City lost about forty acres of wheat.[19] The damage inflicted upon Sunset was not so great. The rains continued for several weeks, threatening destruction of the whole crop, but enough was saved by hard labor to supply breadstuffs for another year. (Fish n.d.a.:7)

Flood waters washed around the new dam at St. Joseph. The dam survived, however, and with some repair was used again in 1878. The harvest in 1877 was recorded as follows:

Sunset:

> 2500 bushels of wheat
> 1250 bushels of corn
> 40 bushels of barley and oats
> 1500 gallons of molasses
>   plus garden and fruit produce
>   "enough to sustain the colony until
>   another harvest". (Warner 1968:14)

St. Joseph:

> 662 bushels of wheat
> 18 bushels of oats
> "some corn". (Porter n.d.a.:355)

Although comparable figures are not available for either Obed or Brigham City, each settlement apparently raised a modest crop (Tanner and Richards 1977:135). However, the settlers at Obed never harvested their 1877 crop, as the settlement was abandoned prior to the completion of the agricultural season that year. The residents of St. Joseph subsequently harvested the crops which had been planted there. The abandonment of Obed, which was situated near a spring that "produced good water and provided abundant meadowland" (Tanner and Richards 1977:136),[20] was precipitated by an epidemic which the settlers referred to as "chills and fevers". While Obed's location near a productive spring offered the potential of high yields, the site turned out to possess serious health limitations. The settlers from Obed scattered among the remaining three towns, with most settling at Brigham City and Sunset (*ibid.*).

While Obed was being abandoned, a new colony was being founded about twenty miles upstream from St. Joseph. A Gentile settler, who had maintained a sheep herd in a 600-acre valley near the confluence of Silver Creek and the Little Colorado River, abandoned his claim late in 1876 (Fish n.d.a.:34).[21] Four men left St. Joseph immediately in order to locate a damsite, survey a ditch, build several log cabins and, thus, lay effective claim to the newly available land. During March of 1877, 9 men settled with their families at the new location and manned the community Woodruff. Fish (*ibid.*) reports that

> this little colony spent the summer and fall in building houses and preparing to put in a dam across the Little Colorado River....The company was too weak to put in the dam this season, and wishing to raise a little breadstuffs went about 20 miles above to Stinsons[22] (sic.) and got permission of him to use what water he did not want. They went to work and put in a small amount of corn and etc. near Stinsons (sic.).

Responding to the small number of individuals who returned to the Little Colorado settlements after the failure of the 1876 harvest and to the desperate need there for increased manpower,[23] church leaders in Salt Lake City dispatched a contingent of recent converts from Georgia and Arkansas to settle among the Little Colorado colonies. Two companies, comprising approximately 100 persons, arrived late in 1877 (Warner 1968:29). As a consequence of this immigration, the ward statistical reports[24] for the four Mormon settlements at the close of 1877 indicate that the lower valley population had been restored to slightly more than the number which had originally entered the region during the spring of 1876 (see Table 2.1). However, the circumstances of these immigrants was described as "destitute" (*ibid.*; see also Peterson 1973:32). While new settlers were needed to furnish enough manpower to launch the 1878 agricultural season, the physical condition of these new immigrants caused them instead to impose a serious drain on the barely adequate harvest of the previous agricultural season.

1878

Even though St. Joseph's dam was not destroyed during the 1877 agricultural season, preparing it for the 1878 season consumed 800 man-days of labor plus an undisclosed amount of labor by teams (Porter n.d.a.:358). Although labor expenditure in the colonies during 1878 had been difficult, rainfall had been sufficient, with the first precipitation occurring on the seasonally early date of May 24 (Jensen n.d.a.:5/19/1878). Brigham City had planted 150 acres in wheat, 50 acres in corn, 12 acres in sugar cane, 4 acres in potatoes, 12 acres in oats, 5 acres in barley, 1 acre in rye, 25 acres in garden and vegetables, and 15 acres in orchard (Nuttall 1878a). St. Joseph reported 100 acres sown in wheat, 40-45 acres

Table 2.1

Population of the Little Colorado Settlements
1877 - 1880

Date	Sunset	Brigham City	St. Joseph	Woodruff	Total
December, 1877	136	277	76	50	539
August, 1878	113	230	64	67	474
September, 1878[a]	102	210	67	65	444
February, 1879	123	235	74	---	---
May, 1879	125	234	65	---	---
August, 1879	124	211	71	---	---
November, 1879	160	241	93	---	---
February, 1880	157	252	96	---	---
May, 1880	150	193	107	---	---
August, 1880	154	151	101	---	---
November, 1880	158	124	103	60	445

SOURCE: Little Colorado Stake, (n.d.).
[a]Population figures for September, 1878 are from Nuttall (n.d.a., n.d.b.).

Table 2.2

Population by Ward
September 1879 and March 1880

Ward	September 1879	March 1880	Percent Increase
Sunset	124	157	26.6
Brigham City	211	252	19.4
St. Joseph	71	96	35.2
Snowflake[a]	341	465	36.4
St. Johns	0	199	---

SOURCE: Table 2.1
[a]Taylor was included in the Snowflake Ward at this time.

in corn, plus other crops (Jensen n.d.a.:8/9/1878). Woodruff listed 33 acres sown in wheat, 7 acres in corn, 2.5 acres in sugar cane and 5 acres in vegetables (Nuttall 1878b), while Sunset is credited with having planted a total of 130 acres in small grains, besides what the settlers there assisted resident Indians in planting (Jensen n.d.a.:7/30/78). Expectations were running high, with a harvest of between 20 and 40 bushels per acre anticipated (*ibid.*).

Then, in late August, an unusually west harvest season was capped by a disastrous flood. Seventeen consecutive days of intermittent rain caused the Little Colorado River to overflow its banks and blanket an area nearly three miles wide to a depth of between one and three feet for about two weeks (Nuttall 1878a; Jensen n.d.a.:9/24/1878). Porter (n.d.a.:358) claims that "on August 22 (1878) the water was a high as driftwood marks showed it to have ever been", and Nuttall (1878b) recorded the following comment during his stay at Sunset:

> During the late flood the water was within a rod from the fort, and was only kept out by a levee on the west side, the wheat field was under water, and much of what was not hauled washed away, and otherwise damaged by the rain, amounted to 500 bushels of wheat, 50 bushels of corn, and 40 bushels of oats and barley.

Brigham City recorded a loss of 40 acres (1,000 bushels) of wheat, 12 acres (300 bushel) of oats, 5 acres (75 bushels) of barley plus 15 bushels of rye, 3.5 acres of potatoes (only saved about 50 bushels) and half the garden and vegetables planted (Nuttall 1878b).

Severe damage was reported for the other towns along the river as well. The dam at Woodruff, which cost that settlement 390 man-days and 227 team-days labor (*ibid.*), was completely washed away, forcing the men of this community to scatter throughout the basin in search of work,[25] with most of them becoming involved in freighting[26] (Jensen n.d.a.:8/23/1878). Some abandoned this townsite altogether, settling in other towns along the Little Colorado or into the new communities forming along the upper reaches of Silver Creek (see below).

Although St. Joseph reported a harvest of 943 bushels of wheat, 40 bushels of barley, 60 bushels of oats and 523 gallons of molasses (Warner 19668:29) (which included an estimated loss of about 275 bushels of wheat alone (Nuttall 1878b), the dam at St. Joseph was claimed to have been in worse condition in August of that year than it was before the settlers began repairs on it the previous February. Furthermore, the river had changed its course so that St. Joseph's dam was now useless. Sunset simply reported that it "had raised a good crop, but much damage had been done to it by the flood" (Little Colorado Stake:8/3/1878). Even the grist mill erected at Brigham City to serve the four settlements was entirely flooded, with its stacks of grain standing 18 inches deep in water (Jensen n.d.a.:8/24/1878). Boats and rafts had to be used at certain locations in order to travel between the various towns (Nuttall 1878b).

Successive crop failures exacerbated morale problems in the four colonies. While some settlements had managed to harvest sufficient grain to maintain their members through the coming winter--if consumed sparingly--Woodruff's crop was a complete failure, and Brigham City had failed to produce an adequate harvest for the third consecutive year. Population figures for the fall of 1878 declined from those of the previous winter,[27] and one community leader reported that "there are hardly a tenth left here of those who were called on this mission, and had it not been for the Lord moving on other faithful men to come here, we would have been in bad condition" (Little Colorado Stake:9/10/1878). Concern over the persistent fluctuation of personnel was a continuing issue among these settlements (*ibid.*:5/6/1878).

Amid mixed harvest yields, 1878 witnessed the inauguration of several new Mormon settlements within the basin. Taylor,[28] the first of these new settlements, was founded on January 22, 1878 by 18 families arriving from southern Utah (Tanner and Richards 1977:36). Located five miles downstream from St. Joseph, Taylor was perhaps the most briefly settled location within the basin; "as the dams built in the Little Colorado washed out, one after another, the people became discouraged and vacated the place in the fall of 1878" (*ibid.*:136-137). During September of 1878, an official expedition under the direction of Erastus Snow, one of the Apostles[29] of the Mormon Church, visited the Little Colorado settlements. L. John Nuttall, secretary and historian for this party, made the following notation regarding the settlement at Taylor:

> When within six miles from St. Joseph, passed the location of the settlement of Taylor, which was on the southwest side of the river. This place was settled in February last by some 15 families. They made a dam and put in their crops, but the bed of the river and banks not being solid, the waters washed the dam away, which caused much extra labor, and the few settlers could not surmount it; in consequence the crops dried up and the people were compelled to vacate the place during the fore part of August, and they have located themselves further up the river (Nuttall 1878a).[30]

Most of the colonists who had abandoned Old Taylor settled along Silver Creek and formed the core of those populations establishing the new communities of Snowflake and Taylor (Fish n.d.a.:33). The founding of Woodruff in late 1877 had marked the initiation of Mormon settlement along this tributary,[31] and during 1878 several additional locations along Silver Creek were opened to Mormon colonization through the purchase of individual non-Mormon claims (Fish n.d.:38-47; McClintock 1921:164-176; Jensen n.d.b.; Peterson 1973:25ff). Erastus Snow's party discovered at least six new settlements on the creek itself plus one, Forest Dale,[32] located beyond the creek "on the Salt River side of the Mogollon Rim" (Peterson 1973:25). Nuttall (1878b, 1878c) counted 175 colonists distributed among these various settlements, ranging in size from 35 at Forest Dale to 22 at Cluff's near Showlow (see also Peterson 1973:25).

Taylor, located about 3 miles south of Snowflake, was first settled in 1877 when settlers from Woodruff obtained permission from James Stinson to use whatever water he did not need (see above). With the failure of the Woodruff dam in 1878, more settlers arrived at Taylor, and a second dam and ditch were constructed. The valley in which Taylor is located is neither as large nor as fertile as that which contains Snowflake.

The final and most significant location settled by Mormons along Silver Creek in 1878 was Stinson's ranch (Fish n.d.:38-42; McClintock 1921:164; Jensen n.d.b.; Peterson 1973:27-32; LeVine 1977:6-17). Irrigating about 300 acres, Stinson claimed all the waters of Silver Creek. With the encouragement of Erastus Snow, William Flake, one of the settlers who abandoned Old Taylor, Negotiated a deal with Stinson to purchase his claim and property for $12,000. Payment was to be made in Utah cattle[33] valued at $11,000, with the balance of the debt settled by Flake and other pioneers harvesting Stinson's crops for him that year. Church headquarters in Salt Lake City subsidized the settlement of this valley by supplying the cattle through which Stinson's property was obtained. In recognition of the central role played by the above-named men, the town was named Snowflake. Snowflake grew rapidly and "quickly became a center of Mormon settlement in Arizona" (Peterson 1973:32).[34]

1879

Eighteen seventy-nine was a relatively prosperous year along the Little Colorado. Sunset had raised 7,000 bushels of wheat and corn (Peterson 1973:19), considerably more than was harvested there for the preceding three years combined. Sunset also reported "much stock" (livestock) for 1879 (Little Colorado Stake:103). St. Joseph reported a "middling good crop not as good as might have been expected, had they been put in earlier" (little Colorado Stake:8/2/1879). Only Brigham City reported poor crops (*ibid*.:103). This first successful season was not to result in local prosperity, however. As news of the good harvest spread back to Utah, additional pioneers were dispatched to settle along the Little Colorado. Because of the favorable reports[35] describing the 1879 harvest, most of the new recruits arrived along the Little Colorado with insufficient supplies to sustain them through the winter.[36] Consequently, much of the 1879 surplus was quickly consumed carrying these extemporized immigrants through what has been described as "one of the worst winters experienced during the entire Little Colorado colonization" (Peterson 1967:41).[37]

The fact that the grain surplus of 1879 was already being stretched thin in order to subsidize new settlements along the upper reaches of the Little Colorado River exacerbated the drain imposed by the immigration of considerable numbers of unequipped settlers. The first and most significant of these new settlements was at St. Johns. As with other suitable locations along the upper river, St. Johns was already settled by non-Mormons and existing claims had to be purchased. Land and water rights at St. Johns were obtained from Solomon Barth for $19,000 payable in Utah cattle (700 head) (McClintock 1921:179; Barth 1973; Peterson 1973:32-34). This amount, almost all of which was forwarded by church headquarters in Salt Lake City, obtained for the Mormon settlers Barth's "squatter's title" to 1200 acres of land and accompanying water rights (McClintock 1921:178).

In order to settle their new location and claim the surrounding lands before the purchase became public, men from all four towns in the lower valley were instructed to locate at St. Johns as quickly as possible (Peterson 1973:34). Strict limits were placed on the amount of land available to any specific individual in an effort to maximize the number of settlers which could be accommodated at the St. Johns site (*ibid.*). By mid-March of 1880, one hundred and ninety-nine Mormons resided at St. Johns, and a call was made in Utah during April of that year for 100 additional families to settle at St. Johns in order to help build this new community. Subsequent calls were made again in 1881 and 1884 to bolster the faltering size of the Mormon population at St. Johns. (see Chapter 7).

1880

By the close of 1880, Mormon settlements had also been established in Round Valley, the present location of Springerville and Eagar,[38] as well as at the mountain locations of Nutrioso, Alpine and Greer (McClintock 1921:184-187; Greenwood 1960:95ff; Peterson 1973:35-36). With the settlement of these latter communities, the expansionary phase of Mormon colonization in the basin had largely come to a close. Mormon immigration into the region had not ceased, however. Continuous reinforcements were needed beyond the turn of the century in order that existing settlements might consolidate their tenuous occupancy of this demanding environment.

The drain of subsidizing successive settlements produced considerable food shortages among donor communities. The loss of grain to support the new settlements until their first harvests (much of which was never repaid), combined with the drain on food resources caused by the heavy influx of destitute immigrants late in 1879, resulted in serious food shortages by the spring of 1880. By June of that year, settlers throughout the basin were reported to be "in almost a starving condition" (Jensen n.d.b.:6/29/1880).[39]

This regional food shortage was made even more acute by the general failure of the 1880 harvest. At Brigham City, crops were "almost an entire failure" (Little Colorado Stake:11/27/1880). Except perhaps for 1877, the settlers at Brigham City had been unable to "raise their own bread" (*ibid.*) during any year, and there was "a feeling among the brethren to remove to some locality where they...(could)...sustain themselves" (*ibid.*). The crops at Sunset were recorded as "very poor" (*ibid.*). While the settlers at St. Joseph had managed to secure a good harvest--including 1,107.5 bushels of wheat, 1100 bushels of corn, 146 bushels of oats, 15 tons of hay, and 8 bushels of lucern (alfalfa) seed (Bushman n.d.)--floods had destroyed their dam before the growing season had passed. Consequently, crops (such as corn) which require a longer growing season failed to mature (*ibid.*:5). So desperate were settlers throughout the basin that a contract was obtained from the Atlantic and Pacific Railroad to construct several miles of roadbed about 150 miles to the east of the region in order to obtain flour and other needed foodstuffs (see Chapter 7).

Colonizing the Little Colorado River Basin with farming communities, thus, proved to be an exceedingly difficult task. Climatic instability, specifically in the form of recurring extremes of drought and flood, and of heat and frost, placed considerable stress on agricultural productivity and on the infrastructure upon which that productivity was based. The costs of community maintenance were high, rendering the establishment of viable communities directly dependent upon such unstable and demanding environmental conditions only limitedly successful. Even St. Joseph, the only original lower valley settlement to survive, experienced near total crop failure during 3 of the 7 years between 1876 and 1882, and as late as 1883 St. Joseph had not produced sufficient flour with which to satisfy its own needs (Little Colorado Stake:3/3/1883).

SUBSEQUENT YEARS

The remaining years of the nineteenth century displayed the same environmental and community instabilities which characterized the 1876-1880 settlement period. While information available for these latter two decades is not as complete as that for the early settlement phase, the picture presented by the available data is quite clear; environmental variability and unpredictability, in conjunction with a limited natural resource base, severely undermined the development of stable and complex communities within the Little Colorado River Basin. This was particularly the case within the lower valley. Environmental variation produced fluctuations in the level of community productivity within specific settlements which, in turn, generated variations in the size of the population associated with the affected towns. Fluctuations in population levels added to the drain imposed on individual settlements by limiting the manpower available to underwrite investments in productive infrastructure--most notably the reconstruction of dams destroyed by recurring environmental excess. Population variability also inhibited the elaboration of diverse productive activities which might have neutralized the impact of environmental instability (see Chapter 6). Systemically reinforced instability was a characteristic feature of Mormon populations and communities in the Little Colorado River Basin until well into the twentieth century.

1881-1900
If the 1880 harvest was "poor", that of 1881 was even worse. St. Joseph harvested only 145 bushels of wheat in 1881, "enough to make flour for 2 months" (Tanner and Richards 1977:78). Due to the time needed to erect a new dam following its destruction the previous season, crops could not be planted early enough to assure

their growth to full maturity. The situation at St. Joseph was apparently so desperate that two men were dispatched from this settlement during September on a two-week trip to Sanders, Arizona (65 miles to the east) in order to secure 3,000 pounds of flour--even though the railroad then under construction was to reach St. Joseph on October 5th of that year (*ibid.*).[40] Brigham City, which had sustained nearly continuous crop failures since its inception, was largely abandoned during 1881, with most of its inhabitants proceeding either to the Silver Creek communities or out of the basin entirely.

Eighteen eighty-two was another year of generally poor harvests throughout the basin. While 1883 was reported as a "fairly good year" at St. Joseph (*ibid.*:45) due to early rains, Sunset suffered the destruction of yet another dam. Farming at Sunset was thus terminated, and the church ceased to call people to settle there (Little Colorado Stake:216G, footnote 75). By 1885, the townsite had been completely abandoned, save for one family.

The next three years--1884, 1885 and 1886--apparently witnessed reduced harvests, and by early 1886 "the people of the Eastern Arizona Stake[41] (were) in poor circumstances. Many saints moved away which caused a decrease in numbers" (Jensen n.d.b.:18866). While the harvest of 1888 was as "good as usual throughout the (Snowflake) Stake" (Fish n.d.a.:66), Fish (n.d.b.:273; quoted in Peterson 1973:143) reported that in 1890:

> The farmers raised but little more than they needed and nothing was sent out in this line for sale, but much that we used in the way of provisions was imported...as there was not enough raised for home consumption....

Eighteen ninety had been a particularly wet year, producing considerable flooding on Silver Creek which destroyed dams at Taylor, Snowflake, Woodruff and St. Joseph. With the exception of Alpine and Heber (New Mexico), the bishops of all the other wards in the St. Johns Stake reported good crops that year (Jensen n.d.d.:12/7/1890). Although the increased rainfall yielded good harvests at St. Johns and other lower settlements during 1890, it resulted in excessive precipitation at higher elevations. At Alpine, "a great deal of grain sprouted in the shock, thereby rendering it almost worthless, and hay that was cut and lay upon the ground rotted" (Jensen n.d.d.:12/7/1890). Generally good crops were reported again throughout the region in 1891, except at St. Johns where heavy losses were inflicted by grasshoppers (*ibid.*:12/6/1891),[42] and at St. Joseph where a flood early in the year destroyed its dam and ditches and washed away considerable farm land (Jensen n.d.c.:2/17/1891).

Between 1892 and 1895, the weather took a decided turn for the worse.[43] The first of these three years was recorded as "exceedingly dry" (*ibid.*:4/1/1893) and the spring of 1893 as particularly windy, "so much so that the whole country seemed to be covered with drifting sands, which made a complete desert of the whole country. Water was very scarce and the grass almost entirely gone" (Jensen n.d.c.:4/1/1893).[44] Cattle died by the thousands (Kennedy 1968:18), as fully one-half of the cattle in Apache County perished (Jensen n.d.d.:4/1/1893). "Stock was nearly ruined and the outlook for all classes of business was very dark indeed (*ibid.*)." Little rain or snow fell again during either 1893 or 1894, and one report at the St. Johns Stake Conference in 1895 (Jensen n.d.d.:3/1895) indicated that many Mormons were in "somewhat destitute circumstances".

Capping a series of devastating years, a combination of high winds, hailstorms, insufficient water (particularly at St. Joseph) and chinch bugs reduced harvests at various settlements throughout the basin in 1895 (Jensen n.d.d.:9/1895). In addition, the entire region endured a killing frost seasonally early in September of that year. Although superior to those of previous years, the harvest of 1895 was insufficient to release many settlers from the accumulated debts they had incurred during the preceding three years (Jensen n.d.d.:9/1895).

While the next three years were comparatively better, 1899 was a different matter; another drought was in the offing. Although the drought of the early 1890's is considered the most devastating to have ever visited the basin, the drought which returned in 1899 was to last 6 years, with relief occurring only during 1903. As early as June of 1899, the following report was presented at the St. Johns Stake Conference (6/12/1899):

> Frosts and cold dissecting winds have destroyed the fruits and damaged or killed much of the grain crops. Crops are nearly a month later that usual. Water is scarce and cattle on the range in some localities are suffering. Cattlemen predict that unless rains come soon many will die. Sheepmen say they have lost much from the cold backward spring and late storms.

VARIATION IN COMMUNITY DEVELOPMENT

For Mormon settlers in the Little Colorado River Basin, the nineteenth century closed under conditions similar to those experienced during the initial years of colonization--that is, with marked fluctuations in agricultural productivity due to environmental extremes. The situation during the intervening years was not much better. Available documents reveal a rather bleak record in which more than half the years under consideration witnessed either poor harvests or total crop failures generally throughout the basin. Although some settlements consistently achieved better than average harvests, others repeatedly sustained inadequate levels of productivity.

The extent of the fluctuations in community productivity experienced by individual settlements during the late nineteenth century is reflected in the annual variation in tithing collected within these same towns.[45] Tables 2.3, 2.4, 2.5 and 2.7 focus on different aspects of the tithing paid during the settlement period and demonstrate that marked variation existed among individual settlements regarding the amount of annual tithing collected.[46] Differences occurred among towns regarding both the average annual amount of tithing collected during the settlement period and the average annual variation associated with the amount of tithing received.[47] Settlements in the basin, thus, differed significantly among themselves regarding both the size and the stability of their annual productivity. Several examples will serve to illustrate this point.

While the mean annual total tithing collected at Snowflake between 1887 and 1905 was nearly 3 times that collected at St. Joseph, the relative annual variation in total tithing received at Snowflake was less than half that received at St. Joseph (see Table 2.3). Similarly, while the average annual tithe paid in field crops at Snowflake was nearly 4 times that paid at St. Joseph, the former displayed only about one-third the relative annual variation in tithing by field crops compared to the latter. Other settlements, such as Showlow and Alpine, collected smaller annual amounts of tithing in field corps than did St. Joseph and experienced even greater relative annual variation in the amount of this tithe (see Table 2.4).[48] Although every settlement in the basin except Alpine and Eagar sustained serious declines in livestock due to the devastating drought of the early 1890's, tithing records suggest that substantial differences existed in the extent to which individual towns suffered livestock losses (see Table 2.5).

Persistent variation in agricultural productivity caused repeated fluctuations in population size within specific towns (see Table 2.6). With each successive crop failure, more individuals left the basin, either returning to Utah or pressing further south in search of more desirable locations in which to settle. Recurring emigration, in turn, stimulated new calls for additional immigrants to bolster faltering communities. However, while all the Little Colorado Mormon settlements experienced population fluctuations, St. Joseph, exhibited over 80% greater annual variation in the relative size of its population than did Snowflake and, considering the two extremes, Alpine registered nearly 3 times the relative annual variation in population size experienced by Taylor.

Although differences in aggregate productivity were, in part, a function of the population size associated with specific settlements, controlling for size differences by converting aggregate tithing into per capita tithing does not significantly reduce the level of variation that existed among these towns (see Table 2.7). While the average annual per capita tithe paid at St. Joseph was about 12% higher than that paid at Snowflake, the relative annual variation in per capita tithe collected at the former settlement was over 70% greater than that received at the latter. At the same time, other settlements, such as Woodruff, Showlow and Alpine, recorded considerably smaller annual per capita tithes than did either Snowflake or St. Joseph and experienced significantly greater relative annual variation in the per capita tithing they received.

Unfortunately, data relating to functional complexity among these various settlements is not nearly as complete as that concerning community productivity or population size. While continuous information on population size and tithing is available for at least a majority of the years under consideration, data on occupational diversity is available only from the 1900 census (see Table 2.8).[49] Information regarding the temporal sequence followed by individual towns in the

Table 2.3

Total Tithing by Ward
1887 - 1905
(in dollar values)

Year	St. Joseph's	Woodruff	Snowflake	Taylor	Showlow	St. John's	Union[a]	Alpine
1887	611	570	2334	1004	364	--	--	--
1888	772	410	2969	1299	283	3637	1427	288
1889	862	602	2855	1573	418	4413	1838	425
1890	1204	797	3411	1710	651	4328	1804	234
1891	560	681	2851	1414	489	4479	1743	688
1892	684	917	3319	1557	522	3447	1391	472
1893	734	581	3865	1130	253	3029	1901	491
1894	583	581	2498	1259	572	2347	1165	218
1895	772	655	2341	995	330	2391	2021	470
1896	748	629	2362	1192	325	2169	1634	442
1897	952	774	2578	1147	389	2541	1975	454
1898	1191	894	2183	1230	226	2861	2156	333
1899	1543	1324	2880	1479	543	3482	2029	594
1900	1141	1394	2945	1563	233	3961	1371	416
1901	1677	1246	3882	1567	--	4200	2485	523
1902	1497	1349	3782	1817	--	3713	1886	391
1903	1907	1578	4142	1857	554	4542	1863	860
1904	1796	1364	3699	2055	585	4186	1739	506
1905	2094	1047	3589	1847	337	4361	2155	583
$\bar{x}$[b]	1124.11	915.42	3025.53	1463.47	430.88	3560.50	180.17	449.33
s	489.41	356.43	589.24	318.20	144.43	817.39	325.69	146.96
V	0.435	0.389	0.195	0.217	0.335	0.230	0.180	0.327

SOURCES: Historical Department, Church of Jesus Christ of Latter-Day Saints (n.d.): Leone (1979).

[a] Union ward consisted primarily of the town of Eagar.

[b] "x" denotes the mean; "s" denotes standard deviation; "V" denotes coefficent of variation.

Table 2.4

Tithing by Field Crops [Grain and Hay] by Ward
1887 - 1900
(in dollar values)

Year	St. Joseph's	Woodruff	Snowflake	Taylor	Showlow	St. John's	Union[a]	Alpine
1887	183	168	876	590	88	---	---	---
1888	271	174	1093	687	69	1223	684	84
1889	424	219	989	617	147	1627	975	93
1890	376	149	1240	493	267	1443	1028	107
1891	97	26	896	658	92	955	934	197
1892	264	239	877	753	139	671	809	140
1893	266	131	955	515	76	673	851	253
1894	250	154	956	604	97	529	353	67
1895	121	237	851	505	89	702	1305	259
1896	254	308	1091	655	141	559	1707	269
1897	268	271	1028	631	241	626	1381	350
1898	262	162	877	771	86	587	1338	256
1899	239	216	898	603	144	730	870	139
1900	205	158	818	532	12	544	153	71
x	248.5	186.57	960.36	617.43	120.64	834.54	952.92	175.77
s	85.12	69.25	117.53	84.52	67.40	366.71	422.36	93.21
V	0.342	0.371	0.122	0.137	0.559	0.439	0.443	0.530

SOURCES: Historical Department, Church of Jesus Christ of Latter-Day Saints (n.d); Leone (1979).

Table 2.5

Tithing by Livestock by Ward
1887 - 1900
(in dollar values)

Year	St. Joseph's	Woodruff	Snowflake	Taylor	Showlow	St. John's	Union	Alpine
1887	229	63	448	39	65	---	---	---
1888	228	84	652	115	33	499	67	96
1889	339	78	593	74	18	329	26	225
1890	243	---	1037	227	137	255	56	52
1891	199	227	730	105	81	269	26	77
1892	161	259	739	165	139	95	37	70
1893	41	136	459	105	98	175	80	74
1894	26	2	356	63	151	126	156	34
1895	64	4	230	37	83	54	48	34
1896	60	2	170	2	33	31	54	84
1897	122	---	195	5	---	54	71	35
1898	112	22	65	6	2	32	206	31
1899	144	11	23	38	8	14	164	149
1900	109	3	14	33	---	95	105	48
x̄	148.36	73.08	407.93	71.29	59.86	156.00	84.31	77.62
s	90.25	90.77	311.86	64.72	54.02	144.58	57.23	55.22
V	0.608	1.242	0.765	0.908	0.903	0.927	0.679	0.711

SOURCES: Historical Department, Church of Jesus Christ of Latter-Day Saints (n.d.); Leone (1979).

Table 2.6

Population by Ward
1876 - 1905

Year	St. Joseph's	Woodruff	Snowflake	Taylor	Showlow	St. John's	Union	Alpine
1876	73	--	--	--	--	--	--	--
1877	76	--	--	--	--	--	--	--
1878	64	--	--	--	--	--	--	--
1879	93	--	--	--	--	--	--	--
1880	103	60	341	--	--	--	--	80
1881	111	54	498	310	--	199	126	161
1882	96	82	340	350	--	358	174	154
1883	101	120	465	354	--	656	218	61
1884	73	164	346	349	--	496	301	77
1885	134	170	379	340	135	503	278	84
1886	151	168	444	243	181	672	301	93
1887	133	174	448	234	125	665	211	192
1888	115	143	363	299	143	615	255	126
1889	107	145	336	315	120	600	248	133
1890	108	178	354	305	149	588	288	88
1891	110	154	295	300	154	569	292	100
1892	107	121	377	279	195	530	329	136
1893	116	143	446	289	179	526	353	138
1894	116	147	450	292	169	460	350	96
1895	128	156	463	300	145	456	379	80
1896	133	168	399	265	154	438	380	81
1897	132	183	407	275	191	438	341	94
1898	101	--	430	283	200	435	352	98
1899	100	198	296	290	124	449	379	125
1900	144	143	312	315	181	436	375	103
1901	156	167	469	318	198	448	305	101
1902	141	181	483	356	215	529	--	99
1903	170	203	440	324	232	509	387	85
1904	179	209	451	340	249	500	375	85
1905	188	199	446	354	275	503	376	83
			469		289	566	379	
X	118	153	404	308	182	506	310	105
s	31.78	40.62	59.68	34.34	47.51	101.70	71.75	31.81
V	.269	.265	.148	.111	.261	.201	.231	.302

SOURCE: Historical Department, Church of Jesus Christ of Latter-Day Saints (n.d.); Leone (1979).

Table 2.7

Per Capita Tithing by Ward
1897 - 1905
(in dollar values)

Year	St. Joseph's	Woodruff	Snowflake	Taylor	Showlow	St. John's	Union	Alpine
1887	4.59	3.28	6.43	4.29	2.55	--	--	--
1888	6.71	2.87	8.84	4.34	2.35	6.06	5.75	2.28
1889	8.06	4.15	8.06	4.99	2.80	7.51	6.38	3.20
1890	11.15	4.48	11.56	5.61	4.23	7.61	6.18	2.66
1891	5.36	4.42	7.56	4.71	2.51	8.45	5.30	3.88
1892	6.39	7.58	7.44	5.58	2.92	6.55	3.94	3.47
1893	6.33	4.06	6.35	3.91	1.56	6.58	5.43	3.56
1894	5.03	3.95	5.40	4.31	3.94	5.15	3.10	2.27
1895	6.03	4.20	5.87	2.49	2.14	5.46	5.31	5.86
1896	5.62	3.74	5.80	4.50	1.70	4.95	4.79	5.46
1897	7.21	4.23	6.00	4.17	1.95	5.84	5.61	4.83
1898	11.79	--	7.38	4.35	1.82	6.37	5.69	3.40
1899	15.41	6.69	9.23	5.10	2.50	7.99	5.41	4.75
1900	7.92	9.75	6.28	4.96	1.18	8.84	4.50	4.04
1901	10.75	7.46	8.04	4.93	--	7.94	--	5.18
1902	10.62	7.45	8.60	5.10	--	7.29	4.87	3.95
1903	11.22	7.77	9.18	5.73	2.22	9.08	4.97	10.12
1904	10.03	6.53	8.29	6.04	2.13	8.32	4.63	5.95
1905	11.14	5.26	7.65	5.22	2.31	7.70	5.69	6.86
X̄	8.49	5.44	7.58	4.75	2.40	7.09	5.15	4.54
s	2.97	1.95	1.54	.80	.77	1.27	.81	1.90
V	.349	.358	.203	.169	.323	.179	.157	.419

SOURCES: Tables 2.3 and 2.6.

Table 2.8

Number of Occupations Listed by Town
1900

Town	Persons Listing Occupation	Occupations Listed
St. Joseph	12	5
Woodruff	31	12
Snowflake	78	16
Taylor	58	8
Showlow	32	7
St. Johns	82	22
Eagar	90	8
Alpine	35	3

SOURCE: Bureau of the Census (1900).
NOTE: This table includes only those respondents who could be distinguished as Mormon for those towns, such as St. Joseph, St. Johns and Showlow, in which a distinct non-Mormon population resided. All individuals listing Utah, Iowa, Illinois, Missouri, Ohio or New York (important states in Mormon history) as the place of birth of persons in the family (depending upon age), were counted as Mormon. Several Mormons were also recognizable by their surnames. Students were not included in this table.

Table 2.9

Establishment Dates for Church Organizations by Ward

Organization	St. Joseph's	Woodruff	Snowflake	Taylor	Showlow	St. John's	Amity[a]	Alpine
Year settled	1876	1877	1878	1878	1878	1880	1879	1879
Ward organized	1878	1880	1878	1880	1878	1880	1880	1880
Sunday School	1876	1878	1880	---	1878	1880	+	---
Woman's Relief Society	1877	1882	1880	---	---	1880	+	---
YMMIA	1883	1883	1880	---	1878	1880	+	---
YMMIA	1883	1883	1880	---	---	1880	+	---
Primary Association	1888	1883	1881	---	---	1882	+	---
School	---	---	1879	---	---	1881	---	---

SOURCES: Jensen (n.d.a., n.d.b., n.d.c., n.d.d.).

[a] Union ward was formed after the dissolution of two previous wards in Round Valley: Amity ward and Omer ward. Amity ward existed between 1882 and 1886 during which time the organizations indicated by a [+] were established (date not specified). Amity ward included those persons resident at the present location of Eagar.

Table 2.10

**Number of Businesses by Town
1905-1906**

Town	No. of Businesses	No. of Business Categories
St. Joseph	3	3
Woodruff	8	6
Snowflake	24	20
Taylor	3	3
Showlow	6	6
St. Johns	34	22
Eagar	20	14
Alpine	2	3

SOURCE: Gazetteer Publishing Co. (1905).

Table 2.11
Rank-Order of Settlements by Specific Criteria

Total Annual Tithing		Per Capita Annual Tithe		Population Size		Businesses 1905-1906		Business Categories 1905-1906		Occupations 1900
X	V	X	V	X	V	#		#		#
St. Johns	Eagar	St. Joseph	Eagar	St. Johns	Taylor	St. Johns		St. Johns		St. Johns
Snowflake	Snowflake	Snowflake	Taylor	Snowflake	Snowflake	Snowflake		Snowflake		Snowflake
Eagar	Taylor	St. Johns	St. Johns	Eagar	St. Johns	Eagar		Eagar		Woodruff
Taylor	St. Johns	Woodruff	Snowflake	Taylor	Eagar	Woodruff		Woodruff[b]		Taylor[d]
St. Joseph	Alpine	Eagar	Showlow	Showlow	Showlow	Showlow		Showlow[b]		Eagar[d]
Woodruff	Showlow	Taylor	St. Joseph	Woodruff	Woodruff	Taylor[a]		Taylor[c]		Showlow
Alpine	Woodruff	Alpine	Woodruff	St. Joseph	St. Joseph	St. Joseph[a]		St. Joseph[c]		St. Joseph
Showlow	St. Joseph	Showlow	Alpine	Alpine	Alpine	Alpine		Alpine[c]		Alpine

SOURCES: Tables 2.3, 2.6, 2.7, 2.8 and 2.10.
NOTE: Settlements are ranked from highest to lowest value on all criteria, except for the coefficients of variation associated with tithing and population size. Coefficients of variation are ranked from lowest to highest.
[a] These settlements ranked equally.
[b] These settlements ranked equally.
[c] These settlements ranked equally.
[d] These settlements ranked equally.

formation of their respective ward organizations is also available (see Table 2.9),[50] as is the number of businesses in each town during 1905-06 (see Table 2.10). Although each of these tables separately can provide very limited information regarding the functional complexity of specific settlements, together they suggest that marked differences existed in the degree to which complex communities evolved at different locations throughout the basin. Regarding both occupational and business diversity and the rapidity and completeness with which they evolved the full complement of church organizations, Snowflake and St. Johns stood apart from the remaining settlements in the basin.[51]

Tables 2.3 through 2.7, thus, illustrate that significant differences existed in both the size and variability of population and productivity among Mormon settlements in the Little Colorado River Basin. At the same time, Tables 2.8 through 2.10 indicate that marked discrepancies also occurred regarding the extent to which functional complexity evolved among these same towns. Table 2.11 suggests that a pattern existed in this variation. With certain notable exceptions, four settlements--Snowflake, St. Johns, Taylor and Eagar--consistently ranked higher than the remaining four towns--St. Joseph, Woodruff, Showlow, and Alpine--regarding the principal indices of community development under investigation.

As part of the explanation of Mormon colonization of the Little Colorado River Basin, this monograph will account for the pattern of variation in population, productivity and functional diversity that existed among early Mormon settlements in the region. The explanation will account for the dichotomy that existed between the two sets of settlements just described and for the regional preeminence of Snowflake and St. Johns. The explanation offered for differences in community development will also be systematically linked to those given for the overall success of the colonization effort, for the role of tithing redistribution in that success and for the differential success of Mormon efforts to establish a viable system of multihabitat resource redistribution. The theoretical basis of this explanation is the ecological model referred to in Chapter 1. Because this model provides the conceptual framework from which local empirical considerations will be examined, it will be described in the following chapter and operationalized to the Little Colorado Mormon settlements. Subsequent chapters will, then, examine empirical developments attending Mormon settlement in the region which bear directly on conditions specified within this model.

NOTES

1. Three settlements--Concho, St. Johns and Springerville--pre-date Mormon colonization of the Little Colorado River Basin.
2. Two significant non-Mormon towns were founded in the early 1880's. Established by the Atlantic and Pacific Railroad, these settlements--Holbrook and Winslow--will be considered only as they affected the development of Mormon towns in the region by

serving as vehicles through which external systems imposed themselves on indigenous developmental processes.

3. The importance of the sense of mission which pervaded Mormon settlement of the Little Colorado River Basin is aptly expressed in the title of Charles Peterson's (1973) history of the local colonization effort and is discussed by him (*ibid*.:38-62). Many colonists, though wishing to vacate the basin, remained until officially relieved of their call by church leaders.

4. Zion is the Mormon appellation for the kingdom of God on earth, and referred in he latter part of the nineteenth century to that broad territory encompassed by Salt Lake City and its numerous widely-distributed satellite communities.

5. In addition to the fact that prior Mormon settlement of the lower valley meant that the better locations along the river had already been taken, the American Colonization Company's decision to leave the Little Colorado River Basin was influenced by the strong anti-Mormon sentiment which pervaded the nation at the time.

6. A portion of the Little Colorado River Basin extends eastward beyond the area under consideration into the state of New Mexico.

7. The term "Saint" applies to any member of the Mormon Church and derives from the proper name of the church, which is The Church of Jesus Christ of Latter-Day Saints. The term "LDS" is also employed by Mormons to refer to themselves, while the name Mormon is a largely externally-imposed label.

8. The four initial settlements experienced name changes soon after their formation. Known initially by the name of the captain of each company, these settlements eventually adopted new names with Mormon significance. St. Joseph experienced an additional name change to Joseph City in 1923 at the request of the U.S. Postal Service. In order to avoid confusion, the four names presented will be used throughout the monograph.

9. Initial Mormon settlement along the Little Colorado River was performed under an organizational arrangement known as the United Order (Arrington 1958:323-349; Peterson 1973:91-122; Arrington, Fox and May 1976; Tanner and Richards 1977:51-63 for discussions of the United Order and of the United Order Movement in Mormon history). Having arisen out of the Panic of 1873, the United Order was in existence among the Latter-Day Saints for several years prior to their settlement along the Little Colorado. The basic principles of the United Order date to the earliest period of Mormon history, and are contained in the Doctrines and Covenants compiled by Joseph Smith, the Mormon prophet. Promoted by church leaders in the late nineteenth century as a method by which Mormon communities could maintain self-sufficiency and independence from the encroaching (and frequently hostile) Gentile society, several settlements were established after 1873 according to United Order guidelines (Arrington, Fox and May 1976:301). The most noted of these settlements was Orderville, Utah (Arrington 1954).

Within the United Order, property and labor were pooled. Upon settlement, individual members turned in their property to the Order and worked and ate as a single body under leaders chosen by them (see for example, the St. Joseph United Order n.d.b.). Labor within the Order was organized along functional lines, with individual interest and experience generally utilized in the allocation of jobs. Little actual specialization of labor existed in such circumstances, however, and the motions passed during United Order meetings display a regular movement of individuals from job to job (St. Joseph United Order n.d.a.). Only those settlements established in the lower valley of the Little Colorado were organized as United Order communities; none of the later settlements founded elsewhere in the basin were organized in this manner. United Orders along the Little Colorado, as well as throughout the church domain, did not persist for very long. By the early 1880's, most communities had abandoned their United Order organization, dispersing properties largely according to family size and need. For St. Joseph, the last colony along the Little Colorado to retain its United Order organization, the final entry into the United Order Minute Book is dated January 29, 1887. The actual dissolution of the United Order at St. Joseph, however, took place much earlier, on November 7, 1882. Much of the aspirations for self-sufficiency and independence within Zion were subsequently transferred to the Co-operative Movement, an integrated productive and marketing arrangement which attempted to incorporate the widely dispersed Mormon towns into a coherent economic system (Arrington 1958:293-322; Peterson 1973:123-153; Arrington, Fox and May 1976:79-110, 311-335; see Chapter 6).

10. Many of the men called were single or, if married, left for the Little Colorado without their families, planning to return for them after the first year's harvest.

11. A conscious effort was made by the settlers at St. Joseph to maintain as accurate an account of their venture as was possible under the circumstances. In addition to "company books", records were kept by some individuals which contain information detailing their daily expenditure of labor and money. Much of this material has survived and has made possible a greater understanding of the settlement of St. Joseph than of any other town in the lower valley (and perhaps in the entire basin). For this reason, greater reference will be made to St. Joseph than to any of the other three initial settlements. However, the similarity of their circumstances minimizes the likelihood that the data available for St. Joseph is not representative of conditions present among the other lower valley settlements. Indeed, where comparable data are available for Obed, Brigham City and Sunset, they tend to reinforce the impression presented by the information pertaining to St. Joseph.

12. Rulon Porter was a past president of the Joseph City Irrigation Company and a son of one of the early pioneers. His writings are, thus, an invaluable source of information concerning the settlement period.

13. At the wage rates employed, St. Joseph's share of the cost of this dam amounted to $2,947.

14. The forts served as the principal residences for the members of these early colonies, even though their defensive function was never needed. Built as a precaution against hostile Indians, each fort was constructed from a similar plan, though not always of the same material. See Peterson (1973:20-21) for a description of these forts.

15. Sand dams were "quickly constructed earthen levees put across the river at low water time following a flood" (Porter n.d.a.:60). The construction of sand dams enabled settlers to obtain water on their land for at least a short time until another flood, or even a slight rise in water levels, destroyed these hastily-constructed structures. While not requiring much initial investment, such dams were susceptible to easy destruction and could not be relied upon as more than temporary solutions.

16. Assuming a ratio of human to team labor similar to that expended on the 1876 dam, 1287 man-days would have been required to build the 1877 irrigation system at St. Joseph, compared to 1208 man-days needed to construct their share of the 1876 water control structures.

17. Dam replacements constituted the single most serious drain on the resources of settlers along the Little Colorado River (see Chapter 5). Tanner and Richards (1977:176) note:

This was especially true of the dam of 1877, which took nearly all of the men in the settlement most of three months to build. There was an additional two miles of ditching to be done, besides plowing and planting and trying to provide homes.

18. The term "lower settlements" refers to those towns located in the lower valley of the Little Colorado River.

19. One source estimated the average yield for wheat during the nineteenth century at about 15 bushels per acre. At this yield, the loss experienced by Brigham City would have amounted to about 600 bushels of wheat. Yields per acre were highly variable, however. In 1878, St. Joseph reported raising 800 bushels of wheat on 95 acres (Jensen n.d.a.:9/26/1878) for an average yield of only 8.4 bushels per acre.

20. Obed Meadow (and Marsh) remained for a long time an oasis in this northern desert. Sustained by springs originating in the Coconino Sandstone (Babcock and Snyder 1947), Obed Meadow remained strikingly lush compared to its immediate environs, at least until heavy pumping of underground water during recent years terminated the surface flow of its sustentative springs (Abruzzi 1985). According to Tanner and Richards (1977:135), "the people of Joseph City...were to keep watch on Obed for a number of years. Later they pastured beef and dairy cattle at Obed, and cut meadow hay and hauled it across the river for a number of years."

21. As indicated earlier in this chapter, Mormons were not the first Anglo settlers to enter the Little Colorado River Basin. In addition to the location of Woodruff, the present sites of Snowflake, Showlow, St. Johns, Concho and Springerville were all occupied prior to Mormon colonization of the region. Such prior occupation of other locations was a principal factor influencing the decision to settle along the lower valley (Peterson 1973:17).

22. Stinson's ranch was located above Woodruff on Silver Creek at the present location of Snowflake. This site, like most others previously occupied, had to be purchased and was obtained from Stinson the following year (see below).

23. Inadequate manpower was a persistent problem among settlements in the lower valley (see below). The intense demand for labor imposed by irrigated farming effectively limited the ability of these settlements to fully and efficiently exploit other resources in the basin which might have enhanced their capacity to sustain losses in the farming sector (see Chapter 6). The following comments recorded in the Little Colorado Stake Minutes regarding Sunset were equally applicable to other settlements in this sub-region:

> Sunset was not very strong in numbers. Their labors were numerous and consequently many industries were neglected that might prove profitable (8/31/1878).

> They were not very systematic in their labors, but did not see how they could better it with so few members, as there was so many things to do and change about with (3/1/1879).

Referring to settlers at Woodruff, which was particularly plagued by persistent labor shortages, Fish (n.d.a.:35) noted that after the 1878 flood, "their scanty numbers compelled them to postpone labor on the dam for the season, and to scatter in different directions in search of work."

24. All of the Little Colorado Mormon towns contained a single ward during the nineteenth century. While St. Joseph and Woodruff remained one-ward towns almost to the present--Joseph City was divided into two wards in 1977--Snowflake has become a six-ward town since forming its second ward in the early 1950's. The fact that each settlement remained a separate ward has facilitated a better understanding of the population growth and economic development associated with each town.

Stake conferences were held four times a year and were generally attended by representatives from each ward. A statistical report for every ward was presented at each conference by these representatives. Serving as important mechanisms for the dissemination of needed information regarding dams, agricultural productivity, health, labor needs and other matters of importance to each settlement (see Chapter 6), these essentially religious rituals included within their agenda the compilation of accurate demographic information for each settlement. Although important deficiencies exist in this data, they do provide a reliable indicator of population distribution among most towns in the region during the latter part of the nineteenth century.

25. One report indicates that only one man and four small boys remained at Woodruff. These had stayed to look after the homes (Jensen n.d.a.:7/30/1878). As had occurred downstream during the initial year of Mormon settlement along the Little Colorado, climatic extremes caused settlers at Woodruff to lose crops to both drought and flooding during the 1878 agricultural season (Nuttall 1878a).

26. Most of the freighting was done to supply Fort Apache which was located to the south of the basin.

27. Between December, 1877 and August, 1878 population decline at Sunset, Brigham City and St. Joseph amounted to 17%, 17% and 16% respectively, while that for the entire lower valley equalled about 12% due to population growth at Woodruff at this time. By September of 1878 the loss of population had increased to 24% at both Sunset and Brigham City, and to about 18% for the entire lower valley.

28. The name Taylor has been applied to two settlements within the Little Colorado River Basin. The settlement discussed at this time has subsequently been referred to in Mormon sources as "Old Taylor", with the name Taylor being reserved for the settlement established later in 1878 about three miles south of Snowflake.

29. "Apostle" is the term applied to any member of the Quorum of Twelve within the administrative hierarchy of the Mormon Church. Thomas O'Dea (1957:178-179) describes the role which this body and its individual members play within the church hierarchy.

> The members of this body elect the First President of the church, choosing their senior member (a precedent started by Brigham Young). The Apostles travel throughout the church, handle administrative business, and comprise a kind of executive high council. (parentheses in the original)

See O'Dea (1957:174-185) for a detailed description of the administrative organization of the Mormon Church.

30. According to William Flake, one of the settlers at Taylor, 5 dams were constructed within 5 months by the residents of that settlement during 1878, all of which washed away (Nuttall 1878a; Fish n.d.a.:33).

31. The settlers at Woodruff did not initially employ the waters of Silver Creek to satisfy their irrigation needs. Due to the expense involved in constructing an elaborate system of

wooden flumes needed to transport this water across steep canyons and in carving ditches along canyon walls, the settlers at Woodruff resorted to irrigation with water from Silver Creek only when the continued loss of dams in the Little Colorado made such expenditures of labor appear worthwhile (Chapter 5). Because the settlers at Woodruff exploited the Little Colorado River for irrigation purposes, the history of this settlement more closely resembles that of the other towns in the lower valley than it does of settlements along Silver Creek.

32. Forest Dale did not survive for very long as a Mormon settlement, as its location turned out to be within the Apache reservation (Fish n.d.:45; McClintock 1921:170-173). Founded on February 18, 1878, the initial colonists were assured by the Indian agent at San Carlos that the town site was not within the reservation boundary which, he stated, was located 3 miles south of the settlement. Nuttall reported that the community was quite prosperous at the time of his arrival and that the settlers there had erected 13 houses, dug several wells, and had 180 acres under cultivation (Peterson 1973:26). In an attempt to placate Indian opposition to their settlement, the Mormons obtained permission to allow approximately a dozen Indian families to join their community. However, this plan backfired when reports circulated that Forest Dale was, indeed, located within the reservation. The Mormon population at Forest Dale quickly dwindled, with most of the settlers proceeding on to the Gila River Valley (the location of Phoenix and Mesa). By December of 1897 only 3 families were left at the site, and these left the following year. In 1881, the rumors shifted; Forest Dale was now claimed to be outside the reservation, and 4 families settled there. Twenty additional families followed, mainly from Brigham City which was gradually being abandoned at this time (see below). In 1882, however, one Forest Dale resident was killed by Indians and another was wounded, and by the following year the settlement was completely abandoned. "The crops were disposed of at Fort Apache and the spring of 1883 found Forest Dale deserted, houses, fences, corrals, and every improvement left behind" (McClintock 1921:173). These settlers also emigrated primarily to the Gila River Valley.

33. Utah cattle served on several occasions as the currency by which Mormons purchased pre-existing land and water rights within the Little Colorado River Basin. A Utah cow (with calf included) was considered more desirable than the indigenous Mexican variety.

34. Snowflake's rapid growth resulted, in part, through the emigration of settlers from other colonies in the basin, particularly those which were in a state of decline along the lower valley. The immigrants from Georgia and Arkansas who settled in the lower valley towns during 1877 were by 1878 "reduced to near starvation" (Peterson 1973:32). Many of these settlers moved to Snowflake. The founding of Snowflake was largely made possible by material subsidies obtained from the lower Valley settlements. Those towns, most notably Sunset, furnished important breadstuffs and seed wheat to the new colony even though they were suffering from less than ample supplies themselves (Little Colorado Stake:3/1/1879).

35. Much hyperbole was contained in letters sent by early settlers to friends, relatives and newspapers in Utah describing life and conditions in the Little Colorado communities in an effort to encourage immigration and, thus, enhance the success of the colonizing effort. This strategy backfired following the harvest of 1789 and produced an immigration which seriously threatened the survival of all who lived in these settlements.

36. As the problem of ill-prepared immigrants became gravely acute, church authorities throughout Utah published announcements directing all prospective settlers along the Little

37. Immigration during 1879-1880 was particularly heavy. Population figures for each of the three lower valley settlements--Sunset, Brigham City and St. Joseph--show dramatic increases between September, 1879 and March, 1880 in spite of the fact that a considerable number of individuals were being siphoned from these same towns to found a new Mormon settlement at St. Johns (see below) and to increase the population of Snowflake and Taylor by over 35% (see Table 2.2).

38. Mormon colonists in Round Valley were initially unable to consolidate into a single ward, due to their acquisition of only scattered landholdings. Later, in 1888, after experiencing difficulties with the nonMormon population at Springerville, a townsite was established two miles south of that town at the present location of Eagar. Several such paired-communities exist throughout the basin--one Mormon, the other predominantly non-Mormon. This has been a prominent feature in the historical development of the region (Duncan 1972).

39. Several accounts exist in diaries dating to this period of persons forced by necessity to scavenge the local countryside in search of various roots, weeds and other less desirable food sources (see also Chapter 7:footnote 4).

40. Two months later, the settlers at St. Joseph received by rail an additional 11,000 pounds of flour, 8,250 pounds of corn and 4,020 pounds of oats (Tanner and Richards 1977:78; see also Chapter 7:footnote 8).

41. The Eastern Arizona Stake was formed in 1879 to accommodate the increasing number of Mormon settlements being established throughout The Little Colorado River Basin. This stake contained all the settlements not located in the lower river valley. Another stake reorganization occurred in 1887. At this time, all the wards located in what is now Apache County were included in the newly formed St. Johns Stake, while those located in present-day Navajo County--including those in the lower river valley--were re-organized into the Snowflake Stake. As a result of rapid population growth within the basin during recent decades, three additional stakes (Winslow, Showlow and Taylor) have been formed out of the Snowflake Stake.

42. McClintock (1921:191) reports that grasshoppers frequently invaded the region between 1884 and 1891 and that Alpine and St. Johns suffered most from the devastation wrought by these insect infestations.

43. During the early 1890's, the Little Colorado River Basin was subjected to a devastating drought which caused widespread crop and livestock losses, and which was to obtrude repercussions upon the basin beyond the period of its immediate duration. The drought of the early 1890's and the circumstances connected with it have been largely responsible for the serious deterioration of grassland communities throughout the region (see Chapter 4). According to Porter (n.d.b.:5), cattlemen invested in large herds following a series of wet years during the late 1880's. As a result, thousands of cattle died (Kennedy 1968:ff.) and the range was destroyed. The potential for substantial livestock losses and severe grassland deterioration which exists when local ranchers stock the range according to the previous year's climatic conditions is illustrated by Porter (n.d.b.:5).

> After a wet growing season herds may go into the winter season with a good supply of range forage, and even though the cold may be severe, they may

survive until another season. But after a dry summer, there may be little forage; and should a severe winter follow, losses may be appalling, possibly as high as 40% or 50%.

44. Heavy winds and their associated dust storms occur regularly at lower elevations in the region during the spring. An erosive factor of imposing dimensions, these intense winds are capable of inflicting severe damage upon soils and crops alike, particularly following years of low precipitation. They were often the object of bitter complaints by early pioneers.

45. Tithing was collected and recorded by each ward before being forwarded to the local stake office and then shipped to church headquarters in Salt Lake City. Most tithing during the nineteenth century was paid in kind due to the scarcity of cash on this isolated frontier. Consequently, items were tithed according to their availability, and tithing stocks were most abundant in the fall following the harvest. For this reason, the amount of total tithing collected by individual wards reflects the aggregate agricultural productivity of each settlement. See Chapter 6 for a detailed discussion of tithing among the Little Colorado settlements.

46. Due to differences in the availability of information, Tables 2.3, 2.6 and 2.7 extend for an additional 5 years beyond Tables 2.4 and 2.5. This additional information was included in the former tables in order to give them greater depth.

47. The information available from Mormon sources on the amount of tithing collected was recorded in dollar values rather than in the quantity of items submitted. Distortions resulting from price inflation are not likely to pose a problem for the period under consideration, however. Through the formation of a regional board of trade (see Chapter 6), the local church organization was able to effectively establish uniform prices throughout the settlement period. Such fixed prices prevailed particularly in transactions with the church itself, including credits given for tithing donations.

48. To the degree that tithing by field crops was directly related to gross farming productivity, Table 2.4 most closely reflects the differential success achieved by individual settlements in their pursuit of the nineteenth century Mormon ideal of a stable, self-sufficient farming community.

49. Comparable data on occupational diversity from the 1890 census was lost when these records were destroyed by fire. Although occupational data can be obtained from the 1880 census, several of the towns investigated here had barely been established by that date, rendering comparisons for that year of questionable value.

50. Including the development of church organizations as a consideration in the evolution of complex communities among Mormon settlements in the region derives from a recognition of the central administrative role performed by the church organization during the nineteenth century (see footnote 9; see also Chapter 6). Because the local and regional church organization provided the near-exclusive governmental apparatus through which the temporal affairs of these frontier settlements were administered--the local church leadership performed legislative, executive and judicial functions and presided over such matters as land distribution, the construction and maintenance of irrigation systems, the settlement of property disputes, the distribution of community surpluses, and the punishment of socially unacceptable behaviors such as thievery and adultery--differences in the complexity of church organization within individual settlements reflect, albeit imperfectly, differences in the administrative resources available to meet local needs.

A ward comprises the smallest administrative unit within the Mormon Church. In keeping with church philosophy, the organization of the ward serves to channel local and church resources into the continued support of the Mormon community. Each ward contains a hierarchical priesthood organization that incorporates all of the males in the ward other than small children. A ward is presided over by a bishop and two counselors who supervise all ward organizations and activities. This complete organization emerges coincident with the formation of the ward. Through time, however, and as resources permit, auxiliary organizations are established within a ward which serve to direct the labors largely of those not included within the priesthood and the council and which perform social tasks in addition to those handled by these initial organizations.

The Sunday School conducts religious education and includes males and females of all ages. The Women's Relief Society provides family, maternal and child welfare services. The Young Men's and Young Women's Mutual Improvement Association (YMMIA and YWMIA) and the Primary Association conduct extensive educational and recreational programs for the community's youth. The YMMIA and YWMIA work with adolescents and young adults, while the Primary Association focuses on small children. Each of these organizations demands the time and labor of the younger and older adults that participate and cannot function when these resources are not available. See O'Dea (1957:180-182) for a discussion of ward organization.

51. The fact that the two stake organizations in the basin were established at Snowflake and St. Johns underscores the regional pre-eminence achieved by these towns.

CHAPTER 3

The Evolution of Ecological Communities

Ecology is the study of living organisms in relation to their environment. Ecological analysis occurs at a variety of levels, including the individual organism, the population, the community, and the ecosystem. Each level encompasses the preceding level and is a component of the level which follows. Thus, populations contain individual organisms and are, in turn, parts of larger communities. The present chapter will focus on the ecological community and describe the theory which explains its evolution. The goal of the chapter is to provide the analytical framework for using general ecological theory to explain developments attending Mormon colonization in the Little Colorado River Basin.

An ecological community comprises all the populations within a prescribed territory, and the major concern of ecological analysis at the community level is with those processes which determine local differences in functional complexity.[1] The evolution of complex ecological communities is the organizational process whereby a growing population adapts to changing conditions of resource availability created in part by its own growth (see Margalef 1968; E. Odum 1971:251-264; Whittaker 1975). Based on principles of energy maximization[2] within systems subject to natural selection, the ecological theory of community development provides a set of general principles from which all community organizational characteristics can potentially be explained. This chapter will offer a general ecological model as a paradigm for explaining the evolution of complex human communities. To achieve this goal, the following discussion will be divided into three parts. The ecological model will first be described as it applies to multispecies communities, because it was in relation to these communities that the general model was initially developed. The model will then be discussed in relation to the evolution of complex human communities generally. Finally, the model will be operationalized to Mormon settlements in the Little Colorado River Basin in order to generate specific predictions regarding empirical developments attending the settlement process. These specific developments and their relationship to the predictions generated by the ecological model will then be examined in subsequent chapters.

THE EVOLUTION OF MULTISPECIES COMMUNITIES

The niche concept is central to explaining the evolution of complex ecological communities.[3] From an energetics perspective, the niche is a function performed within an ecological community which facilitates the flow of resources among that community's constituent organisms. If an ecosystem may be considered a "material path followed by energy" (Margalef 1968:14), then each niche constitutes an individual channel through which energy flows, and a complete sequence of niches comprises an energy circuit. We may distinguish between a population's *fundamental niche*, the exploitative position occupied by that population in a given territory in the absence of competition, and its *realized niche*, that portion of the fundamental niche filled by the population in a particular community containing a specific set of competing populations (see Vandermeer 1972:110-111). The importance of the niche in community evolution lies in the fact that ecological communities evolve as a result of natural selection operating on individual organisms and populations in the community according to their relative abilities to effectively compete for limited resources.[4]

Competition is, therefore, the principal agent determining niche breadth among populations in ecological communities. Where two populations are complete competitors and one is dominant over the entire niche, the less efficient competitor will be completely eliminated from the arena of competition (see Gause 1934; Hardin 1960). If, on the other hand, two populations vary in their relative competitiveness in different parts of the niche, each may evolve to occupy a more restricted portion of is fundamental niche when the two populations occur in the same community (cf. Crombie 1947; Brown and Wilson 1956). As additional populations enter the competition, species specialization and niche differentiation advance, and each population comes to occupy an increasingly reduced portion of its fundamental niche (Vandermeer 1972). Such resource partitioning among competing populations is central to the evolution of complex ecological communities, and an explanation of community evolution rests upon an understanding of the conditions which either facilitate or inhibit this process (see Figure 3.1).

Natural Selection and Community Evolution

Niche differentiation in ecological communities derives from the competitive advantage of specialization. Species which exploit a limited set of resources tend to be more efficient in obtaining those resources than those which must exploit a broad range of resources for their survival (see Levins 1968:10-38; Vandermeer 1972:114-116). Consequently, as additional species enter the community, niches are "squeezed", and the range of resources exploited by individual populations is reduced. The evolution of complex multispecies communities is, thus, an opportunistic process through which natural selection generates the greatest species diversity possible within the limits imposed by resource availability.

Because the community serves as the principal source of selective pressures imposed upon its constituent populations, community evolution must be viewed as a mutual-causal process. While speciation and species replacement contribute to the evolution of complex ecological communities, these same pro-

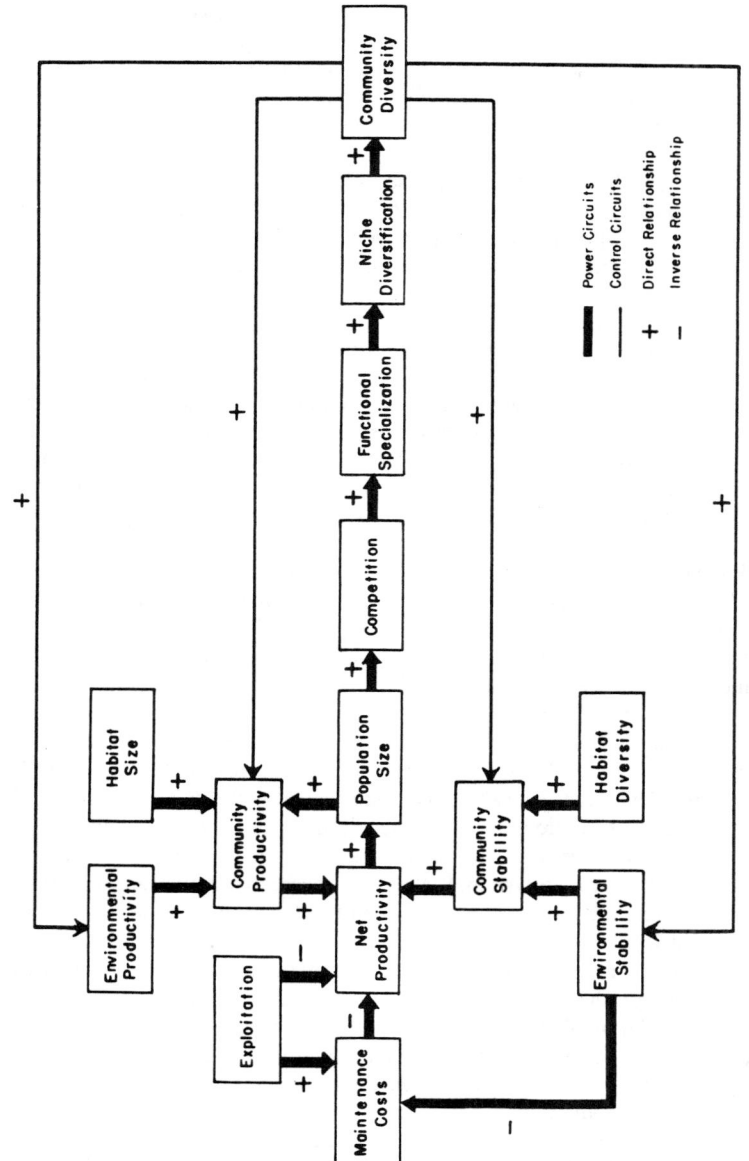

Figure 3.1
The Ecological Model

cesses serve as mechanisms for "adapting parts (populations) to evolving systems" (H. Odum 1971:159). Many species, for example, are adapted to a specific range within the total successional continuum. As the community evolves, the conditions affecting the survival of specific species populations and adaptive strategies in that community change. Furthermore, due primarily to differences in trophic position,[5] particular species are added to the community at different rates. Carnivores, for example, will display a lag in their response to increases in primary productivity compared to herbivores. At the same time, while individuals are added to communities at a constant rate, progressively rarer species are added at a decreasing rate (cf. Sanders 1968), and the ratio of consumer to producer species increases.

Subsidies and Drains in Community Evolution

All ecological communities exist because abundant supplies of potential energy are available in their environments, and their complexity increases to the extent that this potential energy can be converted into community productivity, biomass[6] and, ultimately, functional diversity. However, due to transformational limitations, only a small fraction of the available potential energy can be utilized by a community.[7] A critical variable affecting the amount of potential energy converted to productivity is the cost of maintaining organisms within the community. Due to the maintenance costs of organisms at each trophic level, a continuing reduction in net productivity occurs as energy flows from one trophic level to the next, and a trophic pyramid results within a community in which progressively fewer organisms and species populations exist at increasingly higher trophic levels.

Since maintenance costs constitute the principal factor reducing the amount of energy transferred between trophic levels, they directly affect the amount of biomass and, thus, the level of species diversity that can be supported within a multispecies community. A decrease in maintenance costs increases net productivity at each trophic level, facilitates biomass increase and niche differentiation, and enhances the viability of more marginal niches. It, therefore, increases community diversity. Conversely, an increase in maintenance costs reduces the total amount of energy flowing through the system. It, therefore, decreases supportable biomass, inhibits niche differentiation, reduces the viability of marginal niches, increases the likelihood of local extinction and, ultimately, reduces community diversity.

Any energy source which reduces community maintenance costs increases the amount of energy that can be converted to productivity. It, therefore, serves as an *energy subsidy* to the development of that community. Conversely, any stress that increases community maintenance costs diverts energy away from the community and constitutes an *energy drain*. Energy drains reduce the amount of energy converted to production and, thus, available to support niche diversification.[8] Energy subsidies in ecological communities may include tidal action, rainfall, or enhanced fertilization, while temperature extremes, overgrazing and industrial pollution comprise some energy drains. A specific factor may serve as either a subsidy to or a drain upon ecological systems, however, depending upon other conditions present. For example, while evapotranspiration generally provides a subsidy in humid environments, the same process constitutes a drain in arid ecosystems.[9] Because the evolution of complex ecological communities ultimately depends on resource availability, all phenomena affecting productivity and energy flow in ecological systems may be viewed within an energy subsidy/energy drain perspective.[10] However, research has shown that certain environmental conditions most significantly influence the evolution of complex ecological communities. These include: (1) environmental productivity, (2) environmental stability, (3) habitat size, (4) habitat diversity, and (5) exploitation. Each of these conditions may act independently or in combination to either advance or constrain the evolution of complex ecological communities.

Environmental Productivity. An increase in environmental productivity means that resources which were formerly insufficient to support a particular species population (or adaptive specialization) may now permit that population to survive. As resources become more abundant, specialists outcompete generalists, the niche breadth of populations declines, and community diversity increases. Conversely, a reduction in environmental productivity decreases the competitive advantage of specialization and reduces community diversity. Consequently, organisms and populations in unproductive environments must exploit a wider range of resources than those in more productive habitats. The generally high species diversity in tropical communities compared with arctic and temperate communities derives in large part from the typically higher productivity associated with tropical ecosystems (cf. Rosenzweig 1968, 1976). Similarly, species diversity is generally lower in communities at higher elevations at the same latitude, because organisms in such communities must occupy broader niches (including a wider range of elevations) due to lower environmental productivity (cf. Terborgh 1971).

The relationship between environmental productivity and community diversity is by no means simple and direct, however, and may be compromised by several factors. A key consideration is the rate of energy flow. Organic pollution, which contains high productivity, imposes a stress on ecological communities and reduces community diversity (cf. Patrick, Hohn and Wallace 1954). The decline in diversity which results from organic pollution derives partially from the fact that efficiency is inversely related to the rate of energy conversion in ecological systems (Odum and Pinkerton 1955). In addition, populations at different trophic levels cannot equally capitalize on the enhanced abundance of a specific resource. Consequently, conditions which produce a sudden increase in environmental productivity frequently yield a decrease in community diversity. For these reasons, polluted, rapidly changing and "new" environments (see Slobodkin and Sanders 1969) all act to inhibit the evolution of a complex community organization.

The direct association between environmental productivity and community diversity is also compromised by the existence of specific limiting factors, such as the deficiency of oxygen which characterizes many highly productive eutrophic lakes (see Sanders 1968:267).[11] In addition, random accidents may negatively affect species diversity, even in highly productive communities, by increasing the probability of the elimination of more marginal niches (MacArthur 1972:95; Rosenzweig 1976:129-130).

Environmental Stability. Environmental stability appears to be the single most important factor determining community diversity, and unstable environmental conditions can substantially mitigate, even negate, the positive effect that high environmental productivity has on community development (cf. Sanders 1968; Slobodkin and Sanders 1969; MacArthur 1972; May 1973; Leigh 1976). Where environmental instability prevails to such an extent that substantial resources must be expended simply to maintain or replace existing biomass, little energy remains to support increasing specialization and niche diversification. Moreover, since a population's coefficient of variation has been identified as the principal variable determining its probability of extinction (Leigh 1975), oscillations in the environment increase the likelihood of local species extinction. While resource abundance and reliability permit an increase in biomass, species specialization and niche differentiation, fluctuations in resource availability reduce the viability of marginal adaptations and reverse the effect that competition has on niche differentiation and species packing. Consequently, species specialization serves as a reliable indicator of community stability (Leigh 1975:56).[12]

Environmental fluctuations may vary in amplitude, frequency and predictability. Extreme temperature oscillations in arctic ecosystems produce high maintenance costs and result in the low species diversity characteristic of polar communities. An energy current which is frequently punctuated by extended intervals of reduced flow is likewise incapable of supporting significant species diversity. An increase in the frequency of fluctuations reduces the time available for the evolution of complex energy circuits. In his comparative examination of species diversity in benthic communities,[13] Sanders (1968) established that increased diversity was consistently associated with reduced seasonality.

The most significant dimension of environmental variation is predictability. Natural selection and the economics of consumer choice, which together generate the evolution of complex ecological communities, depend ultimately on a predictable environment. Stereotyped behaviors, which are energetically cheaper and more common in complex communities (Margalef 1968:85), require a predictable and consistent selective regime. The importance of predictability in environmental relations is expressed by Slobodkin and Sanders (1969:85-86). Even though an environment oscillates, they maintain, if it

> fluctuates in a regular and predictable way and with reasonably short periodicity, it is possible for organisms to relate to it by adaptations of very much the same sort as those that occur in a constant environment....

Species diversity seems lower in situations with irregular fluctuations of environmental properties than in structures characterized by regular and predictable fluctuations of the same magnitude.

Because a high degree of specialization can only evolve with reference to a highly predictable environment, the most diverse multispecies communities occur in highly predictable environments with low variability. Thus, the greater diversity of tropical ecosystems apparently derives less from abundant productivity than from environmental stability. Indeed, Sanders (1968) observed that species diversity was not only greater in more stable benthic communities within climatic zones, but also that it was greater in communities located in stable temperate ecosystems than in unstable tropical ones. Significantly, the most complex community observed by Sanders was the shallow water community in the Bay of Bengal, a productive *and* stable tropical benthic ecosystem.

Habitat Diversity. Environments differ in the degree to which resources are evenly distributed and may vary from having resources uniformly spaced (i.e., fine-grained) to their being patchily distributed (i.e., coarse-grained) (see Levins 1968:10-38; Vandermeer 1972:114-116). By providing a coarse-grained distribution of resources, habitat diversity increases the likelihood of niche differentiation and, thus, species diversity due to the greater efficiency of specialized resource exploitation in coarse-grained environments. Several studies have linked species diversity to environmental heterogeneity, including Pianka's (1967) study of lizard species diversity in North America and MacArthur and MacArthur's (1961) analysis of the diversity of bird species in tropical habitats.

Habitat Size. To the extent that environmental diversity is related to the area encompassed by a community, an increase in habitat size is also related to community diversity. Furthermore, the energetic advantage of resource specialization in coarse-grained environments only exist to the extent that the resources provided by differentiated habitats are sufficient to support particular populations and adaptive specializations. Such conditions more likely exist in a larger habitat.

Exploitation. Exploitation occurs whenever one ecological system serves as an energy subsidy for the maintenance or growth of another system. Exploitation imposes an energy drain on the system being exploited. A predator exploits its prey and a herbivore exploits green plants. In both situations, the energy exchanged between organisms is unequal, and one system benefits at the other's expense. Although human populations are the most significant cause of the exploitation of multispecies communities, wherever one arbitrarily draws boundaries in nature an asymmetrical exchange of energy likely occurs across that boundary which increases the organizational difference between the systems concerned (see Margalef 1968). The exploitation of multispecies communities halts their development, which can only resume after the exploitation has been discontinued.

Regulation in Ecological Communities

Because stability increases the efficiency of resource exploitation, natural selection favors those mechanisms which inhibit the occurrence of resource fluctuations in a community. Extended time-lags in the flow of energy between species populations is a major cause of population instability in multispecies communities. Endogenous rhythms nullify time-lags and, thus, increase community stability.[14] Consequently, a selective advantage exists for enhancing the control of endogenous rhythms which render ecological communities increasingly independent of immediate, short-term fluctuations in their environment.

Regulating mechanisms in ecological communities may be divided into *power circuits* and *control circuits* (H. Odum 1971:94). Power circuits are the major channels of energy flow which largely determine a community's organizational structure (for example, where oak trees process 50% of a forest community's energy budget). Control circuits yield only minor energy flows, but are capable of affecting energy flow in the substantially larger power circuits (for example, when the gathering and planting activities of squirrels influence the size of an oak population). A critical change in either the size or the behavior of the squirrel population could have a significant impact on the future position of oak trees in a forest community.

Control circuits are particularly important for the work-gate functions they perform (see H. Odum 1971:38, 44-45), in which one energy flow is enhanced by the multiplicative effect of a supplementary energy input. Agricultural practices such as weeding, plowing and irrigation perform work-gate functions in that they augment the flow of energy that becomes stored in consumable plant material. Increasing stability in ecological systems derives in large part from a greater redundancy of work-gate functions and from the potential that this redundancy offers for circumventing variable energy flows within power circuits.

The greater redundancy present within complex multispecies communities derives largely from the role performed by competing species populations as "compensating devices" (Whittaker and Woodwell 1972:151). Such interspecific competition serves to maintain community diversity. Conditions which eliminate one species (say chestnut) from a forest community may result in another species (such as oak) replacing it in the forest canopy, with the larger community retaining existing levels of productivity, biomass and functional diversity. In addition, interspecific competition reduces the probability of closely related populations exceeding their resource supply, because the size of a particular species population is unlikely to increase dramatically in a community where several formidable competitors provide a constraining force restricting the boundaries of its differentiated niche. Several examples exist of dramatic population increases among species released from the pressure of competition (cf. Russo 1964; Hornocker 1970).

Predation regulates species diversity as well. By influencing prey population size, predation regulates interspecific competition among prey species. Consequently, predation sustains the diversity of prey populations and their resources and, therefore, of the entire community. When predators capable of preventing resource monopolies by individual prey species have been either missing or removed experimentally, the communities involved have become less diverse (see Paine 1966). Thus, while species diversity at lower trophic levels contributes to species diversity in the higher trophic categories (through the flow of energy in power circuits), species diversity at the higher trophic levels can have a regulative impact on the size and diversity of species populations in the lower trophic categories as well (through energy flow in control circuits).

However, diversity alone does not enhance community stability. Indeed, precisely the opposite may be the case. The key to maintaining community stability under variable environmental conditions lies in the degree to which *redundancy* exists in the flow of energy/resources through a community. In multispecies communities, multiple, density-dependent links must exist among species populations at various trophic levels if population regulation within communities is to be effective. Complexity enhances stability in multispecies communities *only* to the extent that species interactions furnish redundancy in community resource flows and, thus, minimize the probability of disruptions in community energy flow in the event that one or more power circuits malfunction. Only where redundancy exists can one population's response to environmental variation be neutralized by the reaction of competing populations, as well as by populations at different trophic levels. Where sufficient redundancy is absent, the negative consequences of environmental fluctuations are likely to ramify throughout the community and reduce community stability, even among communities containing high diversity (see May 1973; Holling 1973; Leigh 1975).

Time is also an important constraint upon the regulative capacity of complex ecological communities, because time is needed to evolve the complex regulative mechanisms associated with diverse communities, and because the evolution of endogenous rhythms requires a stable and predictable environment with the consistent selective pressures that such conditions provide. The control exerted by predators on the size and diversity of prey populations is ultimately dependent on the reliability of the same prey species as resources throughout the year. Thus, the enhanced community stability which results from the regulative effect of community diversity derives ultimately from the productivity and stability of the encompassing ecosystem, because the complex regulative functions performed within ecological communities require continuous and substantial resource flows for their maintenance. Consequently, while capable of mitigating minor disturbances caused by environmental instability, complex ecological communities are particularly vulnerable to major disruptions in the flow of energy, because these undermine the selective advantage of specialization and, thus, jeopardize the niche differentiation upon which their limited regulative capacity is based.

In summation, then, complex multispecies communities evolve as a result of the increasing specialization and niche differentiation generated by interspecific competition. Through increasing intensification of resource exploitation, such communities evolve the most diverse species composition possible within the energetic limits of a particular environment. Because the selective advantage of specialization depends on a resource supply that is capable of supporting increasingly marginal adaptations, community diversity is determined by community productivity. At the same time, since diversity is ultimately a function of net productivity, maintenance costs impose a major constraint on community evolution. Consequently, diverse ecological communities evolve in those ecosystems which support specialized adaptations and which reduce community maintenance costs. These conditions are best met in environments that are both productive and stable, that contain numerous, large and diverse habitats, and that are free from exploitation by other communities. With increasing diversity, ecological communities evolve greater self-regulation and, thus, limited independence from minor environmental fluctuations, provided resource flows within the community possess sufficient redundancy to compensate for local fluctuations in resource availability. However, the intense energy requirements needed to maintain complex ecological communities render these systems particularly vulnerable to major disruptions in their resource supply.

THE EVOLUTION OF HUMAN COMMUNITIES

Like multispecies communities, more complex human communities evolve in large part as a result of the opportunity costs (i.e., selective advantage) associated with increasing specialization, community productivity, and population size. Likewise, development of human communities occurs in relation to resource availability and is directly influenced by environmental conditions which act as subsidies to or drains upon the developmental process. In addition, more complex human communities evolve their own endogenous rhythms which provide them with increasing independence from local environmental variation.

The Niche and Human Communities

As already indicated, the niche is a function which facilitates resource distribution within an ecological community. While species diversity denotes the number of distinct functions in multispecies communities, *occupational categories* and *functional units* define the complexity of resource partitioning in human communities. Although the physical forms presented by species populations and by human occupations and organizations are clearly distinct, each delineates the configuration of productive functions performed within its respective community and, therefore, constitutes an empirical variant of an Operational Taxonomic Unit (OTU) within niche theory (see Vandermeer 1972). Each, likewise, varies in its dimensions as a result of the same competitive process and in relation to resource availability (cf. Clark, *et al*. 1964).[15]

Occupational categories may be defined in terms of the type of activity performed together with the range of resources processed, and may include food production, food distribution, building construction, mining, teaching, and so forth. The precise delineation of occupations within a community is arbitrary, as any recognized occupational category may be divided into increasingly restricted operations. Food distribution, for example, can be partitioned into food transportation, wholesaling, retailing, marketing and processing, while building construction can be divided into carpentry, plumbing, masonry and other construction operations. Even these more restricted operations could be partitioned further if the analysis warranted. Indeed, such increasing specialization of productive functions is central to the evolution of complex human communities.

A functional unit may be defined as any distinct organizational entity that participates in external exchange relations and, thus, facilitates the flow of resources within a community. Among the Little Colorado Mormon Settlements, as in most other recent Western communities (cf. Thomas 1960; Gibson and Reeves 1970; Smith 1976), the functional unit was normally a business establishment. However, functional units as diverse as a communal village organization, a church, an irrigation company and a post office participated in the evolution of Mormon communities in this basin during the settlement period.

Robert Carneiro (1967) has distinguished between "growth" and "development" in human communities. While growth refers to an increase in the number of taxonomic units in a community, development denotes an increase in the kinds of units present. An increase just in the number of farms in an agricultural community would constitute growth. On the other hand, the emergence of new functional units and of occupations other than farming would be an indication of development.

Occupations and functional units, like species, may be arranged into a trophic hierarchy of producers and consumers. This hierarchy is implicit in the economic classification of primary, secondary and tertiary industries and in the distinction between basic and non-basic employment. Within any community, some resource flows may be classified as autotrophic in that they generate the primary resources upon which the remainder of the community depends. Although farming provided the basic community productivity among Little Colorado Mormon settlements, autotrophic functions are not restricted to primary industries in human communities. Both secondary and tertiary industries may serve as the source of basic employment within a particular community. The trophic hierarchy within a specific community must be determined empirically and within a diachronic framework, as local communities may have originated or evolved to exploit a variety of resource bases.[16] Heterotrophic functions distribute the net productivity provided by autotrophic functions throughout the remainder of the community. They may also perform work-gate functions which regulate the productivity of primary producers (see below). Such trophic levels are, of course, abstractions, and actual functional units may operate on more than one trophic level (see Ehrlich and Birch 1964). Just as phytoplankton in northern Sweden alternate seasonally between autotrophic and heterotrophic functions (Rodhe 1955), so also may a food producing unit (such as a farm) in a human community both produce and distribute the food that it grows.

Resource Partitioning in Human Communities

As already indicated, complex human communities, like their nonhuman counterparts, evolve as a result of the opportunity costs associated with specialization among potential competitors. Because the shifting of resources from one productive activity to another involves certain costs, individuals and functional units acquire an adaptive advantage from specialization: both competition and maintenance costs are reduced. By increasing the efficiency of resource exploitation and, thus, the amount of net productivity available for exchange, increased specialization enhances aggregate resource flows within a community (see Samuelson 1958:653).[17] Since certain strategic considerations underlie all social behaviors (cf. Barth 1966; Belshaw 1967; Schneider 1974), the implications of opportunity costs accompanying functional specialization apply to substantively non-economic activities and functional units as well (such as the various organizations associated with local Mormon wards). These must also compete for the limited resources available within a community.

Other things being equal, ecological theory suggests that an increase in community productivity leads to an increase in population size within human communities, because more resources exist upon which additional individuals can be supported. Population increase, in turn, fosters an increase in the number and diversity of occupations and functional units that derive their existence from individual allocations of resources in productive activities. Being opportunistic systems--at least with regard to resource exploitation, functional specialization and community diversification--human communities likewise evolve to the limits imposed by available resources.

Mutual causality operates in the evolution of human communities as well. While occupational and functional unit specialization and differentiation contribute to increasing community diversity, existing productive and distributive arrangements select for the viability of specific additional activities within a community, as well as for general new axes of community development. Moreover, because specific occupations and functional units require distinct population and resource thresholds in order to exist within a community, various functions are added to human communities at different rates during the course of community development (cf. Thomas 1960; Carneiro 1962, 1968; Haggett 1966).

However, an important distinction exists between human and nonhuman ecological communities. Although human communities may, like other ecological

The Evolution of Ecological Communities 67

communities, evolve in response to initial increases in productivity, more often, it would appear, the evolution of complex human communities occurs in response to the adaptive pressures that result from population growth within a fixed habitat (cf. Boserup 1965; Wilkinson 1973; Cohen 1977; Simon 1977; Sanders and Nichols 1988). An increase in population numbers stimulates increases in community productivity and functional diversity by increasing both the supply of and demand for increased resource availability within the community. Permanent increases in population size can only occur in conjunction with increases in community productivity, and population increase within a circumscribed habitat demands an additional intensification of resource exploitation in order to raise the aggregate productivity of a given territory. Pressure for the intensification of resource exploitation places a premium on the specialization of community functions due to the more effective resource exploitation and enhanced net productivity that such specialization provides. Furthermore, due to the Law of Diminishing Returns, population growth within a fixed habitat demands an increase in per capita energy flows which, in turn, increases aggregate community productivity (cf. Boserup 1965:41-55; Harris 1977:176, *passim*).

Continued population growth within a fixed habitat also selects for the evolution of regulative functions that assure sufficient and continued productivity. Consequently, while providing a greater number and diversity of functional units through its effect on productivity and, thus, on the aggregate *supply* of resources in a community, population growth also stimulates, through the increased *demand* for resources that such growth creates, the evolution of functional activities and organizations that, serving as control circuits performing work-gate functions, direct resources into channels expanding community productivity.

Whether specific human communities evolve in response to initial increases in productivity or population growth, the basis of community evolution is the same. The selective advantage of specialization and niche differentiation in either case derives from the opportunity costs that accompany resource partitioning in the presence of an expanded flow of resources. In both situations, the degree to which functional specialization can proceed depends on the viability of individuals subsisting upon increasingly narrow and, thus, more marginal resource flows. Community diversity, therefore, remains a function of the aggregate flow of resources in a community. The greater positive feedback that exists between productivity, population growth and community diversity in human communities than in multispecies communities--rendering such neoMalthusian concepts as carrying capacity not strictly applicable to human populations--also does not undermine the application of the ecological model to human communities. The evolution of human communities occurs in accordance with principles specified within the model. Any population increase associated with the evolution of complex human communities has been founded on an increase in productivity made possible through the evolution of control circuits circumventing environmental limitations (see below). As predicted by the ecological model, increasing community diversity within human communities evolves as a function of concurrent increases in community productivity and population size within the limits imposed by local and regional environmental conditions.

Subsidies and Drains in Human Communities

As with all ecological systems, the maintenance and survival of human communities depends fundamentally on the supply of resources obtained from sources outside the community. The specific external conditions that effect human resource exploitation may also be viewed within a general energy subsidy/energy drain perspective. Moreover, the classification of subsidies and drains in human communities derives from the same criteria employed for multispecies communities and depends on the cost/benefit ratio sustained. Subsidies and drains in human communities likewise assume a variety of empirical forms, and phenomena that serve as energy subsidies under one set of circumstances may act as energy drains under a different set of variable relations, even within the same community. For example, while rainfall, windpower and a nearby stream generally provide relatively cheap energy inputs into agricultural productivity, excess rainfall accompanied by high winds and flooding rivers can severely reduce agricultural production. Likewise, while a reservoir impounded behind a dam provides an important subsidy to agricultural productivity in a climatically variable environment, if the prevailing climatic variation yields excessive precipitation and stream flow, the water stored in the reservoir suddenly poses a dangerous threat.

Finally, the rate at which conditions impose themselves relative to the adaptive capacity of local populations is as important a feature of the subsidy/drain dichotomy in human communities as it is in nonhuman ones. As just mentioned, different precipitation and streamflow rates can have distinct effects on the maintenance costs associated with irrigation and, thus, agricultural productivity in the same farming community. Likewise, while inputs which foster stable population growth actually stimulate the evolution of more complex human communities (Boserup 1965; Culbertson 1971; Wilkinson 1973; Simon 1977), those conditions which yield rapid increases in the size of a community's population--most notably through immigration--are likely to impose a stress on local resources and lead to a greater proportion of these resources being channeled into strictly maintenance functions.[18]

Productivity and Stability in Human Communities. Large discrepancies are unlikely to occur between productivity and biomass in nonhuman communities, due to the Malthusian basis of population ecology in such communities. However, because nonMalthusian principles underlie the evolution of complex human communities, substantial differences in per capita productivity and standard of living occur which complicate the relationship between community productivity, population size and functional diversity in these communities (see Culbertson 1971:35-101; Wilkinson 1973). Because per capita productivity reflects the level of net productivity within human communities, this variable constitutes a necessary supplement to aggregate productivity as a measure of the surplus resources available to maintain community diversity. The evolution of complex human communities, with their enhanced differentiation, interdependence, organization and managerial functions, demands an expensive commitment of community resources and, therefore, depends fundamentally on increases in per capita productivity (see Harris 1959, 1980:183-206; H. Odum 1971; Simon 1977).

Those factors which reduce per capita productivity, then, inhibit community evolution. With regard to the Little Colorado Mormon settlements, the presence of conditions that limited agricultural productivity or that increased the size of the investment required to sustain existing levels of productivity reduced net productivity and constrained community evolution. In the face of unproductive environmental conditions, diverse communities can only evolve where sufficient energy subsidies are provided to mitigate the impact of environmental limitations. Conversely, where productive environmental conditions exist that allow a simultaneous increase in population size and aggregate per capita productivity, increasingly complex human communities can evolve.

The same opportunity costs that limit specialization and that reduce the viability of marginal adaptations in communities located in unproductive environments restrict diversity in unstable communities as well. A variable environment may reduce aggregate agricultural productivity or, by imposing greater maintenance costs, reduce net productivity. Similarly, variations in the amplitude, frequency and predictability of environmental fluctuations yield distinct effects upon the evolution of complex human communities. Differences in the amount of resources required to rebuild dams or in the frequency of dam reconstruction resulted in an unevenly distributed drain on the resources of early Little Colorado Mormon settlements. More important, however, were the differences which existed in the potential for anticipation and control over environmental instabilities. If the principal limiting factor was a variable and unpredictable growing season, little anticipation or control could be exerted. If, on the other hand, agricultural productivity was limited by variation in surface water flow and suitable dam sites were available, a measure of anticipation and control was possible. While the latter condition provided the opportunity to offset environmental variability and induce stable resource flows, the former did not.

Habitat Size and Diversity. Other things being equal, the amount of arable land within an economically exploitable distance from a specific agricultural settlement directly influences the aggregate productivity, per capita productivity, population size and functional diversity associated with that settlement. At the same time, one of the principal features associated with the evolution of more complex human communities has been the integration of numerous, local resource flows into a single, encompassing resource-flow system (see Sanders 1956; Coe and Flannery 1964; Sanders and Price 1968; Flannery 1972; Struever and Houart 1972). To the extent that local habitats differ in ecologically important considerations, habitat diversity facilitates the evolution of more complex exchange systems and of the specialized productive and distributive functions needed to operate them. Through the integration of resource flows originating in multiple, ecologically independent and distinctive local habitats, a more complex and inclusive resource-flow system may evolve (based on more abundant and reliable levels of productivity) than existed in the same territory prior to such functional integration. In addition, where distinct conditions regarding agricultural productivity are associated with different habitats, the integration of resource flows among settlements situated in several independent habitats may significantly offset the developmental constraints imposed by local habitat limitations and contribute further to the development and maintenance of a viable regional resource-flow system.

Since habitat diversity is likely to be a function of the size of a territory, the larger the area incorporated into a unified resource-flow system, the more likely will local environmental instabilities become manageable. As the total geographic area integrated into a single community increases, the abundance of resources available to offset local disturbances will likely also increase, as will the size of the population sharing the costs of counteracting environmental instability.

Exploitation. Exploitation occurs whenever resources which may be used to increase population, productivity or stability within one community are expropriated from that community and used, instead, to enhance the development of another system. As previously indicated, exploitation is a common feature of the exchange which takes place between ecological systems of unequal complexity, and more complex communities generally exploit the less complex systems around them (Margalef 1968). An expanding frontier between ecological communities results largely from the greater competitive advantage that more complex communities possess in relation to the less complex systems on their periphery. The expansion of the American frontier was no different (cf. Shannon 1945). As this frontier expanded into the Little Colorado River Basin, specific resources were expropriated from local use and a substantial drain was imposed on the indigenous Mormon population. The drain on resources, reduced productivity and decreased population[19] that followed the arrival of the American frontier seriously threatened the success of colonization effort (see Chapter 7).

Regulation within Human Communities
The evolution of complex human communities has invariably been characterized by an increase in the number and specificity of regulative functions (i.e., control circuits). Two general kinds of control circuits may be distinguished in human communities: indirect (consumer) and direct (management) regulative functions. The former include those functions and functional units which, through their effect on the demand for specific resources, regulate the output of a community's producers.[20] Consumer functions affect the opportunity costs associated with specific resource allocations among competing producers, and the proportion of consumer functions providing feedback into productivity has increased with the evolution of more complex human communities.

Of greater significance to the evolution of complex human communities has been the increased control exerted by direct regulative functions. More complex human communities possess a larger proportion of management functions to total community organization than do less complex systems, and direct regulative functions have evolved historically to control an increasing share of community resources. Although governmental functional units have performed the principal management functions in human communities since the emergence of the state, important management functions may be performed by functional units other than those under governmental administration. Among early Mormon settlements in the Little Colorado River Basin, the local church organization and its affiliated institutions performed the various management functions needed to facilitate community development.

The ecological model suggests that more complex human communities possess a greater capacity for responding to environmental disturbances than do less complex communities, and that the former systems are more likely to achieve the endogenous regulation of community parameters. With greater independence from local habitat variability, more complex human communities possess a selective advantage in adapting to unstable environmental conditions. The adaptive advantage of more complex human communities is strongly suggested by the anthropological literature (cf. Sahlins 1961; Sanders and Price 1968; Carneiro 1970; Gall and Saxe 1977), and both population size and productivity within specific local environments exhibit greater diversity and stability in more complex human communities than in less complex ones.

As with non-human ecological communities, it is the greater redundancy of resource flows in complex human communities that underlies their ability to achieve limited internal regulation and community stability. If an entire community depends on a single resource flow, variation in the abundance of that single resource will ramify throughout the community. Under such conditions, community stability is largely a direct function of environmental stability. Stability in human communities can only be achieved where community productivity derives from numerous individual producers that are not subject to the same schedule of variation and where alternative resource flows occur in the event of local environmental disturbances. Thus, increasing the number and diversity of distinct local habitats integrated into a single system of resource redistribution would enhance the adaptive capacity of a complex human community by increasing the number of environmentally independent resource flows available to compensate for productive deficiencies in any single habitat (cf. Coe and Flannery 1964; Sanders and Price 1968). The regulative capacity of complex human communities must be viewed hierarchically as well, as complex human communities are only able to offset deficiencies in local production to the productive limits of the territories they occupy and to the extent that aggregate environmental conditions are productive and stable enough to permit the evolution of the specialized functions which underlie resource redistribution (see Abruzzi 1987).

In summation, the extension of ecological theory to human communities suggests that these communities, like their non-human counterparts, evolve as a result of resource partitioning among potentially competing populations. Due to the nonMalthusian basis of human population ecology, however, human communities may substantially enhance the level of population, productivity and functional diversity achieved within a community by increasing the intensification of resource exploitation well beyond that possible in nonhuman communities. However, the potential for positive feedback between population, productivity and functional diversity in human communities does not contradict the general ecological model; rather, it is a special case operating in accordance with the general principles prescribed by that model. Continued increases in population size and community diversity depend fundamentally on increases in the amount and reliability of community productivity. Moreover, like comparable nonhuman ecological systems, the evolution of human communities is subject to those environmental constraints which limit community productivity and stability and which affect the cost of maintaining community operations. While more complex varieties of human and nonhuman communities alike posses an adaptive advantage due to their greater capacity for limited self-regulation, endogenous rhythms in both types of systems depend on a redundancy of resource flows, because only where such compensating energy flows exist can the expensive, complex functions that regulate each system be maintained in the event of environmental disturbances. Consequently, like their nonhuman counterparts, the organization of complex human communities is highly vulnerable to major disruptions in energy flow.

THE LITTLE COLORADO MORMON SETTLEMENTS

If the ecological model is to prove applicable to Mormon settlements in the Little Colorado River Basin, developments attending the settlement process must conform to predictions derived from that model. Those settlements that were located in the most productive and stable habitats and that experienced the lowest maintenance costs in agricultural production should be the ones with the greatest aggregate and per capita productivity, population size, and community stability. These same settlements should also have been the most functionally diverse. Conversely, the least functionally diverse settlements should have been located in habitats that were both unproductive and unstable and that imposed consistently high maintenance costs in agricultural production. These latter settlements should also have displayed the lowest aggregate and per capita productivity, population size and functional diversity in the region. Finally, to the extent that a viable system of resource redistribution emerged and enhanced the success of the colonization effort, it should have been based on the integration of resource flows originating in numerous, independent and structurally distinct habitats experiencing diverse schedules of environmental variation. Only then could resource redistribution possess the redundancy needed for effective environmental regulation.

If the ecological model is to explain variation in community development during Mormon colonization of the Little Colorado River basin, then specific components of the model must be distinguishable among these early settlements. Comparable and appropriate measures must exist for community productivity, population size, community stability and functional community diversity within each town during the settlement period. Likewise, data must exist which permit a clear comparison of the environmental conditions--including environmental productivity and stability, habitat size and diversity and the extent of exploitation--associated with each town and with the region as a whole. Differences in these environmental conditions must be shown to be related to differences in community productivity, maintenance costs, stability, population size and, ultimately, functional diversity.

Population size can be determined for each Mormon town throughout settlement period from figures available in the Statistical Reports of the Mormon Church. These same reports contain data on the amount of tithing collected by each ward (see Table 2.3). Nearly all tithing collected along the Little Colorado during the nineteenth and early twentieth centuries was paid in kind and according to availability. Consequently, tithing figures closely reflect community productivity. Mean annual tithing collected by the appropriate ward will be used as the measure of average annual *aggregate productivity* for each settlement.

Two independent indices of functional diversity exist for Mormon towns in the region during the settlement period. These include the number of occupations declared in each settlement during the 1900 census and a directory of business establishments located in each town during 1905-06 (see Tables 2.8 and 2.10).[21] A merging of these indices provides a comparable measure of the functional *community diversity* of each town at the end of the settlement period and indicates the relative extent of heterotrophic elaboration achieved.

Data also exist which permit a consideration of the differential subsidies and drains experienced by individual Mormon towns. Per capita productivity can readily be calculated by dividing total tithing by population (see Table 2.7). This figure indicates the *net productivity* achieved by each town independent of population size. *Community stability* may also be calculated from the data already presented. The coefficients of variation for population size, total tithing and per capita tithing provide measures of the stability of population, aggregate productivity and net productivity respectively.

No single variable provides an ordinal ranking of individual settlements regarding environmental productivity and stability, habit size and diversity or exploitation. However, sufficient data does exists which allows a characterization of individual habitats with respect to these conditions. For nineteenth century Mormon agricultural settlements in the Little Colorado River Basin, *environmental productivity* was determined largely by soil quality, the abundance and quality of irrigation water, and by the average length of the growing season. *Environmental stability* was primarily a function of the extent of fluctuations in water supply and in the length of the growing season associated with a particular habitat. *Habitat size* was determined by the amount of arable land contained within the valleys in which individual settlements were located. *Habitat diversity* was determined largely by the specific combination of factors which underlay environmental productivity and stability associated with each settlement. The major source of *exploitation* experienced by the Little Colorado Mormon settlements concerned their interaction with competing nonMormon interests in the region.

If the ecological model is to prove applicable to the Little Colorado Mormon settlements, then those settlements that produced the largest and most stable aggregate tithing, per capita tithing and population size should also have achieved the greatest occupational and business diversity. A rank-ordering of individual settlements according to a composite measure of the conditions underlying the evolution of complex communities--the mean and coefficient of variation associated with total tithing, per capita tithing and population size--should correlate with a rank-ordering of the same settlements according to a composite measure of functional diversity. Settlements ranking highest on both of the above composite indices should have been located in large valleys that were both productive and stable and should have suffered the least from exploitation. With regard to the ability of complex communities to compensate for environmental instability, the model predicts that those Mormon efforts at resource redistribution which were based on a greater redundancy in the flow of resources through the redistribution system should have been the most successful. Furthermore, to the extent that such redundancy was achieved, it should have been able to compensate for drains imposed by either environmental instability or exploitation by external sources.

Eight settlements were chosen for comparison in statistical tables. These settlements were selected largely for considerations of data consistency and availability and because of their subregional preeminence. Several settlements were abandoned during the nineteenth century. Others were unable to maintain independent ward organizations and, therefore, lost their separate statistical reporting. Extinction and statistical merging thus restricted the number of settlements for which comparable data is available. The eight settlements chosen for comparative purposes also possess a greater volume of supporting data, enabling a fuller appreciation of their developmental context. Although considerable information is available on the colonization of the Little Colorado River Basin, due largely to the Mormon concern for accurate record keeping, as with most historical research, the extent of information available for specific settlements is uneven. While a truly remarkable amount of detailed information has been preserved for St. Joseph (see Chapter 2, footnote 11), only scattered references exist regarding the details of daily life in Showlow, Eagar, Alpine and the other more remote highland communities. The availability of information on the remaining settlements falls between these two extremes.

The goal of this book is to provide a synthetic ecological explanation for successful Mormon colonization of the Little Colorado River Basin which systematically and parsimoniously accounts for: (1) local differences in community development, (2) the role of tithing redistribution in successful colonization, (3) the differential success of Mormon efforts to develop a viable multihabitat resource redistribution system, and (4) the impact of external factors on the settlement process. However, before the general ecological model outlined above can be applied to explain these various developments, specific circumstances associated with the colonization effort need to be examined. Since the natural environment presented the principal set of material conditions influencing the settlement process, the analysis will begin with a description of the major physical features of this basin.

NOTES

1. In multispecies communities, the communities most frequently studied by ecologists, complexity has largely been equated with the diversity of species populations (cf. Whittaker 1975; Pielou 1975).
2. Because all applications of resources necessitate an expenditure of energy, the effects of which are determined by thermodynamic principles, ecological communities may be viewed as energy-flow systems, and the terms energy and resource may be used interchangeably. The advantage of an energetics approach to the analysis of ecological communities lies in the simplicity that this approach offers for modeling complex systems and in the potential that it provides for incorporating empirically distinct ecological communities within a unified theoretical perspective (cf. E. Odum 1971:37-85; H. Odum 1971; Little and Morren 1976).
3. The niche encompasses several dimensions of a population's existence that affect its contribution to the flow of resources through a community (see Levins 1968; Vandermeer 1972). These include: (1) the populations's habitat or spatial location, (2) its functional role within the community (including both consumptive and non-consumptive behaviors), and (3) its distribution along environmental gradients.
4. Considerable debate exists among ecologists regarding the level at which selection operates in ecological systems (see Odum 1969; Smith 1964; Williams 1971; Wilson 1980; Engelberg and Boyarsky 1979; Patten and Odum 1981). The position adopted here is: (1) that living systems, including human and nonhuman genes, cells, tissues, organisms, populations and communities, exist within a hierarchy of organized ecological relationships, and (2) that the emergent properties of more inclusive systems select for greater coherence in the activity of subordinate units (see Stebbins 1968; Slobodkin 1968; Leigh 1977; Alexander and Borgia 1978; Abruzzi 1982:28-29).
5. Trophic position refers to the consumptive, energy conversion function performed by organisms within an ecological community. In multispecies communities, trophic levels may be classified as autotrophic (self-nourishing) and heterotrophic (other-nourishing). Autotrophs, such as green plants, convert the sun's energy into organic matter and provide the gross primary productivity upon which the remainder of the community depends. Heterotrophs, which include herbivores, carnivores and decomposers, convert one form of organic matter into another and are dependent on the prior existence of autotrophs (see E. Odum 1971:8-22 for a discussion of the trophic structure of ecological systems).
6. Because multispecies communities contain organisms of differing size, simple population totals provide an inadequate measure of the organic matter maintained within such communities. Total biomass must be calculated instead.

7. Primary productivity constitutes the energy budget of a community. Gross primary productivity, the total rate of photosynthesis less the maintenance of producers (mostly green plants), equals net primary productivity, the amount of energy stored in the form of plant material. Net primary productivity, which is substantially less than gross primary productivity, constitutes the total energy supply available to maintain those populations (herbivores) which feed on the producers.
8. See E. Odum (1971:43-53) for a discussion of energy subsidies and drains in ecological systems.
9. Evapotranspiration is the loss of moisture which results through evaporation from soils and transpiration from vegetation (see E. Odum 1971:123-124).
10. Leibig's "Law of the Minimum" and Shelford's "Law of Tolerance" (see E. Odum 1971:106-136) are general principles regarding material (and, therefore, energy) drains within ecological systems.
11. Eutrophic lakes are shallow, fertile lakes with complete sunlight penetration.
12. The opportunity costs associated with functional specialization limit the diversity of social organization in single-species communities as well. Among social insects, for example, Rapport and Turner (1977:330) report that "a fluctuating environment can make a particular caste uneconomical and favor generalists over specialists even if the functions the caste performs remain as important as before" (see Wilson 1968, 1971).
13. Benthic communities are those communities located at the bottom of aquatic ecosystems.
14. Considerable controversy surrounds the extent to which regulation occurs within complex ecological communities (cf. May 1973; Leigh 1976; Engelberg and Boyarsky 1979; Patten and Odum 1981; Abruzzi 1982, 1987; Moran 1984). MacArthur (1955) and others (cf. Elton 1958; MacArthur and Connell 1966; Margalef 1968; Odum 1969, 1971; Brookhaven National Laboratory 1969) have argued that more complex ecological communities are inherently more stable than less complex communities due to the ability of their intricate food webs to regulate energy flow and, thus, to offset the destabilizing impact of environmental fluctuations. Based on his observations of model ecosystems, May (1973) concluded that no mathematical basis existed for this thesis and that complex systems were, in fact, less stable, although he acknowledged the general empirical association between diversity and stability in ecological communities. He, therefore, proposed that stability promotes complexity.
The important issue is not whether complexity yields stability or whether stability results in complexity, but rather the development of a synthetic model of community development that accommodates the empirical research supporting both of these theses. In fact, the above theses likely reflect processes operating at distinct levels in hierarchically organized ecological systems (see Alexander and Borgia 1978; Abruzzi 1982:28-31, 1987). Furthermore, to say that regulation occurs in ecological systems does not imply the existence of a homeostatic balance--an equilibrium--between populations and resources, as has frequently and mistakenly been suggested (cf. Odum 1969; Margalef 1968; Rappaport 1968; Leone 1979; Patten and Odum 1981). As it is used here, the term regulation refers not to a self-regulated state of affairs but rather to processes through which the activity of one population positively influences the abundance and distribution of resources and, therefore, conditions which affect the existence of other populations in a community.
15. Resource distribution in both human and nonhuman communities actually results from the productive activities of individual actors participating in a variety of independent

exchange relations. Species and comparable functional categories in human communities comprise equivalent Operational Taxonomic Units from the perspective of niche theory in that both delineate the canalization of resource flows in their respective communities. While a precise calculation of all individual productive activities would provide the optimum basis for examining functional diversity and resource partitioning in ecological communities, such specific data is unavailable for most communities. Consequently, more general categories must be used.

16. The precise form and function of specific functional units in human communities evolve from the specific ecological and infrastructural context of the community in which they exist. For this reason, functional units in separate and unrelated human communities are not strictly comparable (no more than are distinct species in separate and unrelated multispecies communities strictly comparable in terms of the dimensions of the niches they occupy). Due to differences in niche breadth and dimensionality that accompany resource partitioning among functional units in distinct and unrelated human communities, such units are unlikely to discharge fully equivalent niche functions. Such functional uniqueness underscores the necessity of analyzing evolutionary processes through the use of diachronic data and calls into question cross-cultural comparisons of social complexity based on trait counts or on the occurrence of specific traits in an overall sequence (cf. Carneiro 1962, 1968).

17. Recognition of the competitive advantage of specialization and of the increased productivity that results from interdependence among differentiated productive units dates to Classical economic theory and was formalized by David Ricardo in his Theory of Competitive Advantage, which forms the basis of contemporary economic theory regarding opportunity costs. As a result of the continuing exchange of models between ecology and economics (see Rapport and Turner 1977), the principles of opportunity costs attending specialization have been successfully extended to explain, among other things, prey specialization among predators (cf. Schoener 1971) and the number and functional diversity of castes among social insects (cf. Wilson 1968).

18. By consuming resources which could otherwise have been invested in productive activities, rapid population growth reduces the flow of net productivity available to support increasingly specialized and, thus, more marginal adaptations. From this perspective, overpopulation may be defined as any excess of population over productivity that reduces per capita resource availability and, thus, threatens the survival of existing organisms in the community. Overpopulation contributes to community simplification because more marginal niches become less viable as pressures mount to channel an increasing proportion of a community's resources into maintenance functions and to broaden the range of resources exploited by individuals and by specific functional units (cf. Lee 1968, 1972; Tanaka 1976).

19. In some cases, colonial exploitation may actually stimulate population increase (cf. White 1973, 1975; Nag, White and Peet 1978). However, because population growth in such circumstances is accompanied by a decline in per capita resource availability--and, thus, in the material base available to generate an indigenous surplus--the complexity that normally accrues from increases in population and productivity cannot be supported.

20. In much the same way that the numaym and the potlatch stimulated community productivity among the Kwakiutl of the American Northwest (cf. Piddocke 1965) and that the "big man" and reciprocal feasting achieved the same result among Kaoka-speaking peoples of Guadalcanal (Hogbin 1964), a successful restaurant or restaurant chain may

stimulate the production of specific food resources within a contemporary Western community.

21. Although information also exists which indicates considerable variation in the extent to which each town developed an elaborate church organization (see Table 2.9), this data does not readily lend itself to operational definition with respect to the ecological model and, therefore, will not be used in the following analysis.

CHAPTER 4

The Little Colorado River Basin

The natural environment clearly presented the primary set of conditions influencing community development among early Mormon agricultural settlements in the Little Colorado River Basin. Local differences in community development were largely associated with the distinct pattern of subsidies and drains imposed by local conditions of resource availability. A consideration of the material conditions influencing community development in this region must, therefore, begin with a detailed description of the physical environment to which these early pioneers had to adapt.

Topography

That portion of the Little Colorado River Basin under consideration encompasses a territory of approximately 5,000 square miles currently sandwiched between the Navajo and Apache Indian reservations to the north and south respectively and bounded on the east by the New Mexico State border and on the west by the Coconino County (Arizona) line. The study area consists, therefore of the non-reservation portions of Navajo and Apache Counties in northeastern Arizona.

The Little Colorado River Basin is located at the southern periphery of the Colorado Plateau which extends over most of the Four Corners Region.[1] The basin appears as an undulating, saucer-like plain sloping to the north and northeast. Rising from an elevation of 5,000 feet in the valley of the Little Colorado River (which forms the northern boundary of the region under consideration), average elevation increases within 75 miles to about 8,000 feet along the Mogollon Rim, a steep escarpment defining the southern boundary of much of the region (see Map 4.1). The topographical slope within the southern basin[2] increases with elevation: while the average slope between Holbrook and Snowflake is 18 feet per mile, the slope between Snowflake and Showlow increases to 39 feet per mile, and that between Showlow and Lakeside (8 miles south of Showlow) increases still further to 72 feet per mile. With slight variation, these gradient figures are representative of the slope at corresponding altitudes throughout most of the two-county region.

The southern highlands within the study area, the White Mountains, contain several peaks exceeding 10,000 feet in elevation, with Baldy Peak, the highest of these, reaching an elevation of 11,403 feet. Baldy Peak, which is located in the southeast corner of the study area, gives rise to the Little Colorado River, the principal stream of the river basin (see below). The Little Colorado River Basin

80 Dam That River!

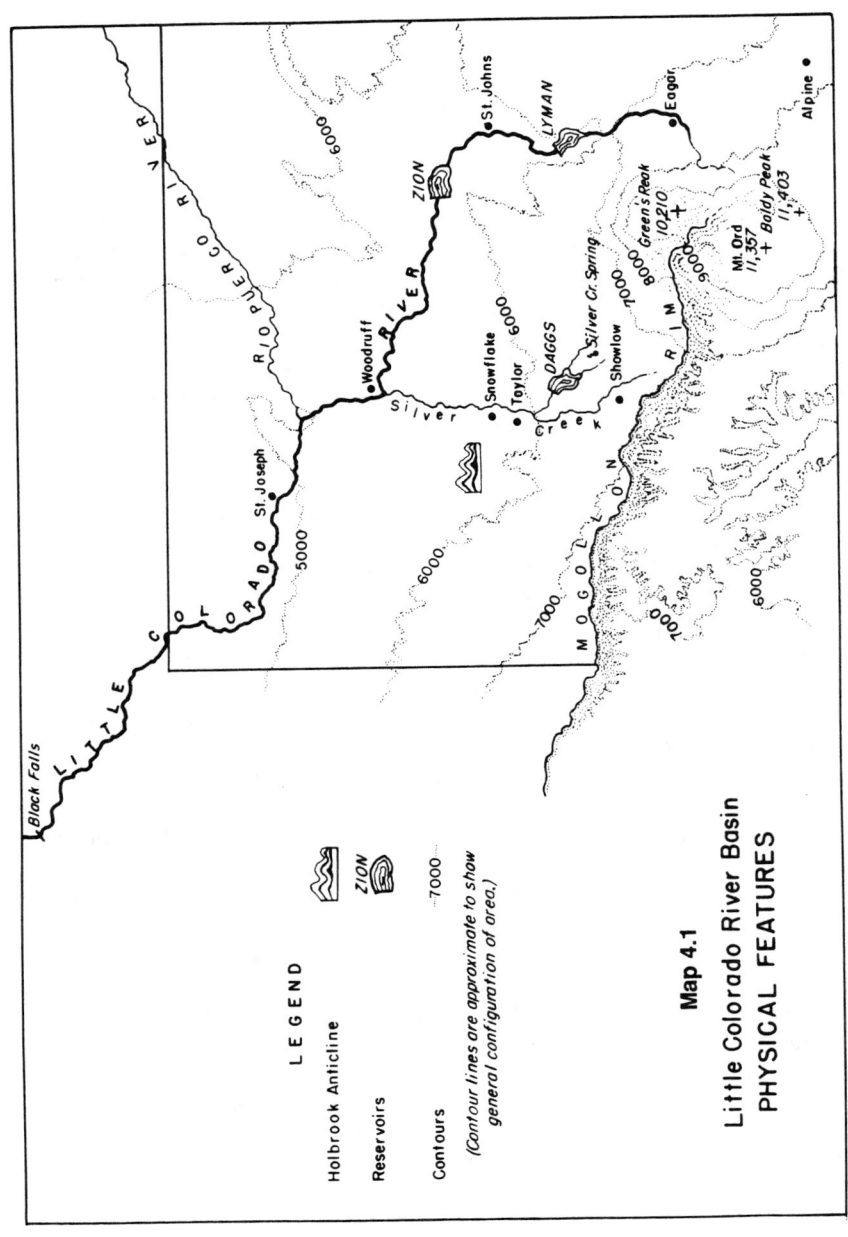

Map 4.1
Little Colorado River Basin
PHYSICAL FEATURES

has been accurately described as "a plateau which has been dissected by a major stream and its tributaries" (Harrell and Eckel 1939:28).

Climate and Weather

Topographical differences have resulted in significant spatial variation in climate throughout the study area, as altitude represents the single most important variable affecting mean annual temperature and precipitation. Because the storms which provide the major sources of precipitation enter the Little Colorado River Basin from the south, the White Mountains act as "orographic triggers" (Jurwitz 1954:12) and receive the lion's share of the moisture which these storms yield. Acting as "domes of cold air" (*ibid*.), the mountains to the south force the moist air in these storm systems to ascend; this rising air cools until the condensation point is reached and precipitation results. Since all of the Little Colorado River Basin is located on the leeward side of the White Mountains (at least with regard to the predominant southerly storm systems), specific locations within the basin receive lesser amounts of precipitation as their elevations decrease and as their distances from the mountains to the south increase (see Table 4.1). Consequently, a close association exists between altitude and precipitation along any particular longitudinal gradient. St. Johns and Springerville, located further to the east than most of the other towns under investigation, deviate from the general trend presented due to the northeastern slope of the region. However, precipitation figures at these two towns are consistent with the relation between altitude and precipitation elsewhere in the region.

Precipitation occurs within the basin primarily during the months of December through March and July through September. For many locations the remaining months may produce little, if any, precipitation. This seasonal variation in precipitation derives from the fact that weather conditions in this region (and throughout most of Arizona) result largely from two distinct storm systems that affect the region at different times of the year.

Winter precipitation is primarily determined by a high pressure system originating in the Pacific Ocean near the Hawaiian Islands.[3] This largely unstable air mass pulsates periodically, transposing its axis between a north-south and an east-west orientation. As a result, it alternates from extending eastward over the Pacific Northwest--and at times even to the Great Basin--to retreating to over 1000 miles from the continent. The recession of this "southwestern" storm systems permits the entrance of "northwestern" storms which deposit lesser amounts of precipitation.

Summer rainfall results primarily from the "Bermuda High" located off the coast of Florida which experiences pulsations similar to those described for the Pacific air mass. Beginning in the latter part of June, the Bermuda air mass moves westward, traveling great distances over warm tropical waters in the Caribbean and in the Gulf of Mexico. As this storm system crosses land, its "moist river of air is exceptionally deep, at times extending to the fifteen and twenty thousand foot levels" (Jurwitz 1954:12). Much of this moisture is released upon contact with the White Mountains.[4]

Table 4.1

Elevation and Precipitation for Selected Locations in the Little Colorado River Basin

Month	Mean Precipitation (in inches)						
	Winslow (4850')	Holbrook (5075')	Snowflake (5582')	St. Johns (5725')	Showlow (6331')	Springerville (6965')	Lakeside (7054')
January	.56	.62	.69	.67	1.78	.50	1.99
February	.40	.61	.84	.57	1.38	.53	1.97
March	.47	.60	.85	.78	1.81	.49	1.98
April	.46	.54	.68	.53	.79	.46	1.37
May	.26	.27	.45	.42	.49	.46	.61
June	.36	.40	.62	.45	.46	.57	.56
July	1.62	1.85	2.39	2.27	2.47	3.23	4.36
August	1.71	1.42	2.57	2.23	2.38	2.84	2.98
September	.78	.94	1.35	1.30	1.73	1.33	1.72
October	.52	.69	.89	.66	1.23	.78	1.78
November	.52	.72	.61	.56	1.41	.62	1.39
December	.93	.59	.68	.75	1.71	.66	1.73
Total	3.59	9.25	12.62	11.19	17.64	12.47	22.44

SOURCE: Harrell and Eckel (1939).
NOTE: Since they are situated in the same valley and at similar elevations, meteorological information collected at Springerville is applicable to Eagar. Close proximity also results in comparable precipitation data for Snowflake and Taylor. Also, because St. Joseph and Woodruff are located near Holbrook, climatological data for Holbrook may be considered representative of that for these and other lower valley settlements.

Winter precipitation has been more significant than summer rainfall in terms of its contribution to agricultural productivity in the region, even though at the lower elevations the relative abundance of summer precipitation increases significantly (see Table 4.2). Summer precipitation generally occurs through brief, spotty and torrential thunderstorms which unleash a flood of water for short periods of time. However, due to the intense nature of these summer storms, soils quickly become saturated and unable to retain much of the moisture they receive. Consequently, most summer precipitation streams down well-worn arroyos, carrying with it soil particles which are then deposited into river channels. Prior to the introduction of farming, most of this silt either settled in riverbeds at lower elevations, or was deposited into the Colorado River. The remarkable scenic beauty of the Painted Desert (which is located along the northern perimeter of the Little Colorado River Valley) derives largely from the erosive impact of these summer storms. However, with the introduction of agriculture into the basin, much of this silt has accumulated in irrigation reservoirs or settled upon agricultural lands. In either case, the extensive silting of streams produced by summer storms has exacerbated the cost of farming in this region (see below).

Table 4.2

Relative Contribution of Winter and Summer Precipitation at Selected Locations in the Little Colorado River Basin

Location	Winter Precipitation (in inches)	Percent of Total	Summer Precipitation (in inches)	Percent of Total
Lakeside	10.43	46.5	12.01	53.5
Springerville	3.26	26.1	9.21	73.9
Showlow	8.88	50.3	8.76	49.7
St. Johns	3.86	34.5	7.33	65.5
Snowflake	4.35	34.5	8.27	65.5
Holbrook	3.68	39.8	5.57	60.2
Winslow	3.34	38.9	5.25	61.1

SOURCE: Table 4.1.

Winter precipitation contributes more to the regional ecosystem than its relative abundance would suggest. While approximately 50% of the annual rainfall over much of Arizona is received during the months of December through March, runoff into reservoir systems due to winter storms accounts for nearly 85% of annual values (Jurwitz 1954:10). Because winter precipitation is deposited more evenly within specific locations; because its moisture is released as runoff more gradually; and because the vegetation is dormant during the winter months and cannot compete with human populations for the water available, the importance of winter precipitation to communities in the Little Colorado River Basin has been paramount. Since the time of their initial colonization, several human settlements and the agricultural systems upon which they have depended have relied fundamentally upon the adequacy of winter precipitation and upon the water storage which it provides.

Growing Season

The length and reliability of the growing season at any specific location is also a function of altitude. Average annual number of frost-free days decreases as altitude increases, as does the variance associated with this mean (see Table 4.3). The average growing season varies from 179 days at Holbrook to only 87 days near Alpine. In addition, significant variation in length of the frost-free period occurs at any specific location from year to year. At Holbrook the length of time between killing frosts ranged from 203 days in 1936 and 1942 to 144 days in 1939, while the number of frost-free days at Alpine ranged from 114 days in 1933 and 1936 to only 36 days in 1944. Although the length of the growing season has varied significantly at Holbrook and the other communities in the lower valley, frosts have not seriously jeopardized agricultural productivity in that portion of the basin.[5] At Alpine, on the other hand, temperature variability has all but precluded significant agricultural enterprise. At intermediate locations within the basin, the effect of temperature variability is not as clear; fluctuations are more likely to crisscross the boundary defining a minimum and reliable growing season. Snowflake, which has an average growing season beyond the minimum length needed for healthy crops, and which has experienced very few years below that minimum, has displayed a frost-free period ranging from 184 days in 1940 to only 86 days in 1936. St. Johns, with an average growing season of 159 days, like Holbrook, has not experienced a single growing season of less than 120 days. The closest that St. Johns approached this figure was 123 days in 1930 and 126 days in 1926 and 1941.[6]

Due to the central importance of altitude in determining local climatic conditions, an inverse relationship exists between precipitation and both the length and reliability of the growing season in the basin. The inability to simultaneously maximize these three prerequisites for successful farming has effectively limited significant local agricultural development to river valleys at lower and intermediate elevations. Generally, elevations above 6,000 feet do not provide sufficiently reliable growing seasons. In addition, climatic variability has yielded crop failures even at locations which display sufficient average growing seasons. Elevations below 6,000 feet, on the other hand, do not receive sufficient precipitation to produce healthy crops. Although dry farming has been attempted (mostly at higher elevations), extreme fluctuations in water availability have made this a largely unsuccessful agricultural strategy in this arid region (Harrell and Eckel 1939:32).

Table 4.3
Mean Annual Number of Frost-Free Days

Location	Years of Record	Elevation (in feet)	Mean Number of Days	Standard Deviation	Coefficient of Variation
Holbrook	29	5,075	179	16.459	.092
St. Johns	25	5,725	159	20.139	.126
Snowflake	23	5,582	144	24.388	.169
Springerville	24	6,965	130	18.220	.140
Alpine	18	8,030	87	18.719	.215

Source: U.S. Weather Bureau (n.d.).

Only in the lower and intermediate river valleys, where adequate growing seasons exist and where a sufficient supply of surface water is available, could viable agricultural communities develop in this arid river basin during the nineteenth century. Irrigated farming at lower elevations has been an agricultural imperative with deep historical roots throughout the Southwest.

Plant Communities

Because local temperature and precipitation are so closely connected with altitude, more or less clear plant community gradients exist throughout the region. Most of the Little Colorado River Basin falls within the Upper Sonoran floral life zone, which at this latitude occurs between 4,000 and 7,000 feet (Akers 1964:9), and four floral communities are recognized in order of ascending elevation: desert, grassland, juniper-piñon woodland and montane forest (see Map 4.2).

Northern desert vegetation consists principally of sage-brush, shadscale and saltsage, and exists mainly in the Little Colorado River Valley along the northern perimeter of the region. Joseph City (St. Joseph) and Woodruff are located within this zone.

The grassland community is the largest single vegetative zone in the basin, encompassing over 40% of the total area (see Table 4.4). This community contains principally gama, ring, muhly, needle, triple awa and galleta grasses (Salt River Project 1974 section 3:42), with various desert shrubs intermingled. Snakeweed is the dominant shrub, representing 77% of all shrubs in the grassland community north of Snowflake (*ibid.*:82). Sagebrush is second in distribution, comprising 20% of the shrubs present in this community. In the eastern portion of the grassland community, near St. Johns, the proportion of snakeweed decreases to 55%, with rabbitbush comprising the remaining 45% (*ibid.*:102).

In the grassland community, vegetation provides only a partial cover. Bare soil accounts for 55-65% of the total surface cover (Dames and Moore 1973 [section 4]:201). Moreover, herbaceous vegetation accounts for only 14% of basal cover in the grassland community near St. Johns (Salt River Project 1974 section 3:102), and an even smaller 3% in the grassland community north of Snowflake (*ibid.*:82). Since grasses comprise the principal food resource for livestock, sustainable animal densities in the grassland community are quite low.[7] In addition, because grasses provide a dense root system which helps to counter erosion pressure during periods of high runoff (e.g. during intense summer storms), the lack of effective floral cover contributes significantly to the intense soil erosion which characterizes the region, and to the heavy silting of the Little Colorado River which this erosion has produced.

Continuing southward, as elevation increases grassland gives way to juniper-piñon woodland, the second largest plant community in the basin. The dominant species within this community are one-seeded juniper, Utah juniper and piñon pine. Throughout most of this community piñon pine is present, but not to the extent of the juniper varieties (Gookin, *et al.* 1972:36). However, the proportion of piñon within the woodland community increases with elevation.

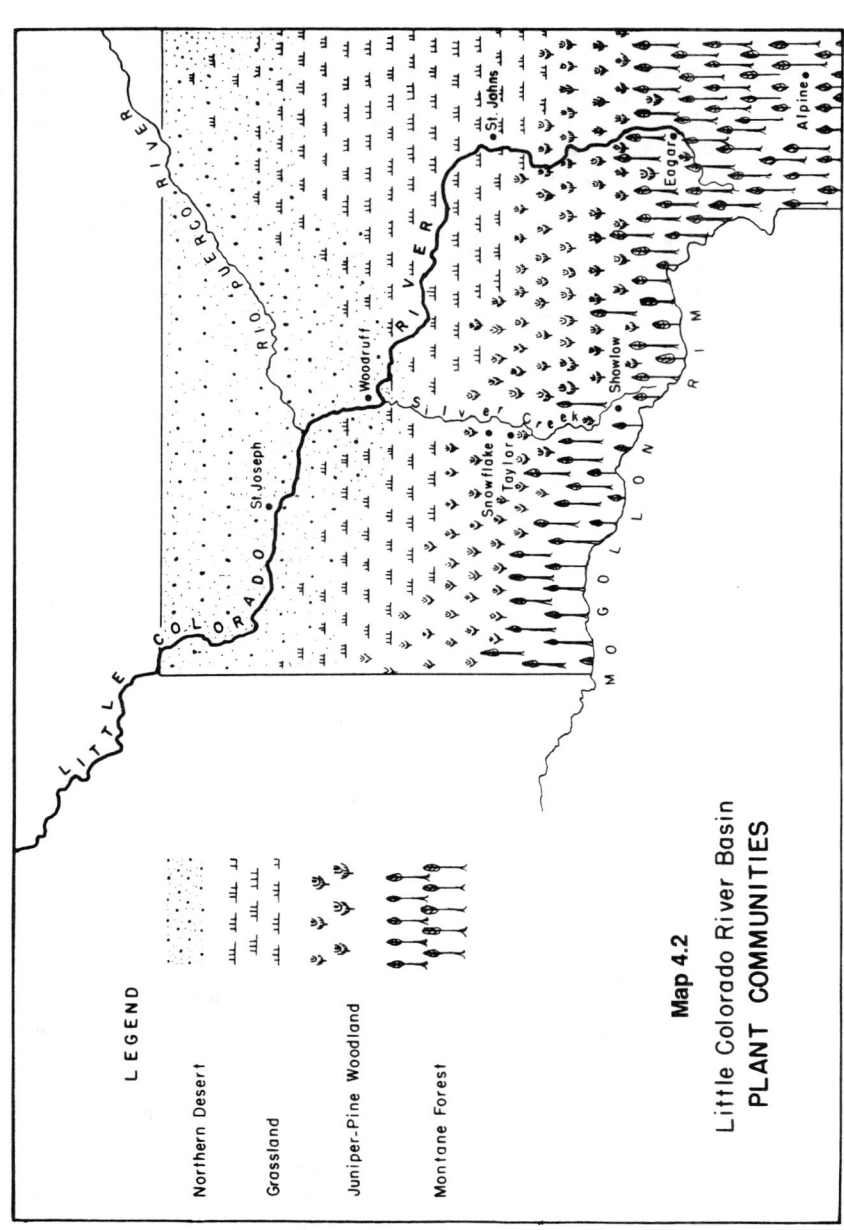

Map 4.2
Little Colorado River Basin
PLANT COMMUNITIES

Both juniper and piñon are short, woody types of vegetation which, due partially to their shallow root systems, dominate in rough, broken country with shallow soils (Harrell and Eckel 1939:27). Density of tree cover increases with precipitation, as does the proportion of piñon in the juniper-piñon woodland community. Thus, both plant density and the relative proportion of piñon increase along a southerly gradient until a transition occurs near 6,000 feet to the montane forest community dominated by ponderosa pine. Due to variation in precipitation, an east-west gradient also exists in woodland density and in the proportion of piñon pine. While tree density in the St. Johns area averages only 10 trees per acre, 30 trees per acre is the mean near Snowflake (Salt River Project 1974 [section 3]:79, 99), which is located at the same approximate latitude and elevation as St. Johns.[8]

Table 4.4

Area and Relative Proportion
of the Four Principle Plant Communities
in the Little Colorado River Basin

Plant Community	Area (in acres)	Percent of Total
Northern Desert	169,200	3.6
Grassland	1,922,027	41.2
Juniper-Piñon Woodland	1,797,004	38.5
Montane Forest	776,401	16.6
Total	4,664,632	99.9

SOURCE: Little Colorado River Plateau Resource Conservation and Development Project (1972).
[a]Total does not equal 100% due to rounding error.

The montane forest community begins as juniper-piñon woodland gives way to ponderosa or yellow pine. The transition from woodland to forest occurs quickly as elevation in the southern highlands increases sharply. Ponderosa pine is the dominant species in this community up to an elevation of about 8,000 feet', after which aspen and Englemann spruce dominate at succeedingly higher elevations. Alpine meadows occur over extensive areas above 9,000 feet.

The ponderosa pine community in this region is part of the largest continuous stand of ponderosa pine in the United States. Beginning north of the Grand Canyon, this forest stretches in a southeasterly arc through the area under investigation and into western New Mexico. Because of its particular suitability for lumber, the local ponderosa pine community is presently the center of an active lumber and wood-related industry. While logging and lumber production have always been important economic activities locally, most of the large-scale lumber-related activities post-date World War II. As late as 1936, 96% of the ponderosa pine in the study area was estimated to be virgin timber. Several towns in the region have depended significantly upon the lumber industry.

The alpine meadows which dot the higher elevations have continuously provided an integral link within the agricultural cycle of communities in the region. These meadows have been employed for the summer grazing of cattle (and previously sheep) at least since the initial settlement of the area by Mormon pioneers. The availability of summer pasture at higher elevations (which is currently regulated by the National Forest Service)[9] ultimately determines the number of livestock which the region can support. Dependence solely on the grassland and juniper-piñon woodland communities would not permit economically viable animal densities.

These four biotic communities account for nearly 100% of the surface area in the Little Colorado River Basin.[10] While the grassland and juniper-piñon woodland communities together account for nearly 80% of the total land area, each of the four zones has had its unique impact on the history and development of communities in the basin.

Soils

The complex geological history of the Little Colorado River Basin has produced a diverse variety of soils in the region (see Kester, *et al.* 1964; Miller and Larsen 1975). Because considerable micro-regional differences exist in soil composition, generalizations have only limited applicability. However, with this caveat in mind, soils within the basin may be grouped according to their physiographic position, including: (1) soils of the flood plains and low alluvial fans, (2) soils of old alluvial fans and terraces, and (3) soils of the uplands (Kester, *et al.* 1964:7). Except for soils in the flood plains (which do not normally extend for more than a mile on either side of the Little Colorado River) and those at higher elevations (within the montane forest community), soils throughout most of the basin are thin and loamy. The productive capabilities for individual crops of the specific soil types discussed are presented in Table 4.5.

The soils of Joseph City (Jocity soils) are primarily alluvial in origin and lie north of the Little Colorado River on nearly level to moderately sloping alluvial fans and river terraces (see Kester, *et al.* 1964:22-23). Originating from the Moencopi formation (Bureau of Reclamation 1950:3), these soils have also received parent material derived from alluvium washed down from the deposits of Chinle shales located in the badlands of the Painted Desert, as well as from beds of fine gravel, sand and loam (Kester, *et al.* 1964:22). Jocity soils are deep, and their profile is characteristically clay loam throughout. Due to the important silt and clay composition of these soils, they provide low permeability, and are thus particularly susceptible to flooding during intense summer precipitation.

Having developed under dry climatic conditions, these soils are also typically low in organic matter and are deficient in nitrogen and phosphorus. Furthermore, since both the Moencopi and the Chinle formations contain large concentrations of alkali salts, soils in the vicinity of Joseph City (St. Joseph), particularly those which have been irrigated, contain sodium in quantities sufficient to inhibit--even preclude--healthy plant growth (Bureau of Reclamation 1950:3-4). The increased salinity and silt deposition caused by irrigation has only aggravated problems inherent in Jocity soils prior to human exploitation (see Abruzzi 1985:261).

Table 4.5

Average Estimated Optimal Yields of Principal Crops
Grown on Selected Soils under Prevailing Management Conditions

Soil Type	Wheat (bushels)	Barley (bushels)	Oats (bushels)	Corn (tons)	Alfalfa (tons)
Bagley	45	50	50	17	6.5
Clovis	60	--	70	23	6.0
Eagar	25	--	36	17	2.0
Jocity	--	45	--	11	5.0
Luth[a]	--	--	--	--	---
Navajo	--	--	--	10	4.0
Showlow	22	--	50	10	3.5

SOURCES: Kester, et al. (1964:61-62); Miller and Larsen (1975:39).
NOTE: No entry indicates that the crop listed is either not suited to the soil or is not commonly grown on it.
[a]No figures are given by Miller and Larsen for any of the crops listed on Luth soils. The only harvest data provided for these soils is for wheat and oats produced for hay. These figures were 1.0 and 1.5 tons per acre respectively.

The predominant soils under cultivation at Woodruff are of the Navajo series. Navajo soils comprise reddish-brown, calcareous clay reaching a depth of more than 60 inches. Located on nearly level to gently sloping flood plains along the Little Colorado River, these soils derive from a parent material consisting of alluvium washed from shale, sandstone, limestone and basalt. Navajo soils are typically low in fertility, moderately well-drained and very slowly permeable. Navajo soils in the vicinity of Woodruff were apparently moderately saline prior to the introduction of irrigation. However, the continued application of clear water

The Little Colorado River Basin

from Silver Creek has leached much of the salts contained within the root zone of most crops (see Kester, *et al.* 1964:28-29).

The soils in the gently sloping floodplains near Snowflake and Taylor are primarily of the Bagley series (see Kester, *et al.* 1964:11-13). Derived mainly from sandstone, limestone, shale, basalt, volcanic cinders, sand and gravel, these soils are well drained. Brown to light brown in color, the surface soil is either clay loam, loam or sandy clay loam in most places. Kester, *et al.* (1964:11) describe these soils as follows:

> The Bagley soils are among the most fertile and productive in the Area. They are well-drained, are moderately permeable, and have a high water-holding capacity. Runoff is very slow to slow. All except Bagley loam have an extremely stable, granular structure in the surface soil and do not erode easily.

Soils in the vicinity of Snowflake and Taylor are also generally free from harmful accumulations of soluble salts even after a century of continuous irrigation. (Bureau of Reclamation 1947:42; Salt River Project 1974 [section 3]:78).

Soils in the vicinity of St. Johns are of the Clovis series (see Miller and Larsen 1975:10-12). These are generally well-drained soils formed in sand and gravelly alluvium derived from quartzite, gneiss, schist, sandstone and limestone. The Clovis soils near St. Johns are sandy clay loam. Runoff in these soils is medium, as is the hazard of erosion. The topsoil, subsoil and underlying material are all generally categorized as of moderate alkalinity (Miller and Larsen 1975:10). Permeability and available water capacity are also rated as moderate (*ibid.*).

Due to greater precipitation and denser vegetation, soils at higher elevations in this region are not characteristic of the arid Southwest. They resemble more closely soils found in the northeastern part of the United States than they do the neighboring soils at lower elevations (Greenwood 1960:39). Soils of the Showlow series are the most extensive soils in the region, and are found at many locations in the montane forest community (see Kester, *et al.* 1964:34-37). Showlow soils are deep, built upon a parent material of alluvium and contain large amounts of sand, gravel and cobbles. These soils are slowly permeable, well-drained and have a good water-holding capacity. Because of their greater organic content, Showlow soils are rated as moderately fertile, and on soils with less than a 3 degree slope the erosion hazard is slight.

The principal soils under cultivation in the vicinity of Eagar are of the Eagar series, which comprise well-drained soils formed in a gravelly alluvium derived from basic tuff (see Miller and Larsen 1975:12-13). The surface layer of Eagar soils is dark grayish-brown and dark gray gravelly loam about 9 inches thick (about 15 inches thick where the slope is less than 3 degrees). Permeability is moderate and the water-holding capacity is generally low in these soils. A temporary water table occurs in some irrigated areas during the growing season due to overirrigation and normal ditch losses. Eagar soils are also moderately alkaline and calcereous throughout. Runoff is medium with a moderate hazard of erosion.

In Bush Valley, the location of Alpine, the primary soils are of the Luth series (see Miller and Larsen 1975:19). The Luth series consists of poorly drained soils that are formed in alluvium derived from basic tuff. These soils are generally found in narrow flood plains within mountain valleys. Their surface layer is typically dark gray clay loam and clay about 18 inches thick. The soils of this series are slightly acid in the surface layer, though mild to moderately alkaline in the underlying material. Permeability is high, as is available water capacity.

A final comment regarding the physiological structure of the basin is warranted. Perhaps the most important geological development within the region, at least from the perspective of establishing viable agricultural communities in the area, was the rise of the Mogollon Geanticline and the subsequent volcanic activity that occurred (see Harrell and Eckel 1939; Babcock and Snyder 1947; Greenwood 1960; Akers 1964; Dames and Moore 1973; and Salt River Project 1974 for discussions of the geological history of the region). This uplift created the higher elevations which form the southern boundary of the basin and which enhance the physical diversity of the region. By creating an "island of humid climate" (Greenwood 1960:19) in an otherwise arid environment, several distinct floral communities were established. More importantly, increased water supplies became available to both surface and sub-surface sources. In addition, the uplift and subsequent erosion produced deep narrow canyons in the southern portion of the basin which made the construction of storage reservoirs possible.[11] The topography of the southern half of the region contrasts markedly with that of the northern half, where few adequate sites exist for the storage of irrigation water. Consequently, settlements in the lower valley of the Little Colorado River had to rely exclusively on diversion dams rather than storage reservoirs as the principal infrastructural supports of their agricultural systems. In contrast to farmers at Snowflake, St. Johns and other Mormon towns to the south, those in the lower valley towns (such as St. Joseph and Woodruff) remained at the mercy of an unpredictable and highly variable river which regularly ceased flowing during a critical period in the agricultural cycle.

WATER

The role that variation in the abundance and distribution of suitable irrigation water played in the development of agricultural communities in this region cannot be over-emphasized. The availability of good-quality water functioned as a critical limiting factor governing the development of stable agricultural communities, and variations in the supply of usable water were closely associated with local fluctuations in population size.

Water presents itself to the basin in three forms: precipitation, surface flow and groundwater, and has provided in each form a distinct impact on community productivity and stability in the basin (Abruzzi 1985). These three water sources are necessarily interrelated. Precipitation is the ultimate source of water supply for the region as riverbeds and underground aquifers depend on rain and snow for their water. However, while the contribution of water in the form of precipitation is largely limited to its immediate area of deposition, that water which finds its way into river channels and into underground streams provides an impact far beyond the limits of its initial ingression. Due to the differential impact and local availability of these three forms of water, each should be treated independently as a distinct condition affecting community development in the region.[12]

Precipitation

The previous discussion has already demonstrated that considerable spatial and temporal variation exists in precipitation throughout the basin. However, two additional comments regarding precipitation warrant consideration at this time. First, the specific pattern of its temporal variation renders precipitation an inadequate primary source of water for the support of farming activities throughout most of the basin. The onset of the summer storms, which usually begin in July, occurs too late to autonomously support the agricultural cycle, because a substantial proportion of the annual irrigation requirements (about 45%) need to be applied to fields during the dry months of April, May and June (see Table 4.6). Thus, much of the growing season transpires during the spring dry season when insufficient precipitation occurs to facilitate the germination and growth of healthy crops.

Secondly, precipitation cannot provide a reliable source upon which to base an agricultural system, because most summer rainfall occurs in the form of scattered, torrential and unpredictable storms. Such precipitation finds its way either into the various river channels or into underground aquifers. Since most of the water which enters the basin as precipitation is ultimately lost to the local system,[13] any attempt to establish agricultural communities dependent primarily upon precipitation would be doomed to failure.

Surface Water

Surface-water flow throughout the basin, being a function of precipitation and ambient air temperature, follows a generally regular annual cycle (see Figure 4.1). Due to the existence of substantial snowpacks in the upper watershed (see Table 4.7), intermittent warming intervals, combined with precipitation, produce moderate runoff in most river channels during the months of January, February and March. Gradually, this flow subsides as the snowpacks disappear, and by late May and early June most streambeds throughout the basin are dry.[14] With the onset of summer storms in July, the volume and velocity of water flow in the river beds increases dramatically.[15] The passing of summer storms and the consequent decline in precipitation throughout the region causes runoff to subside again until snow re-accumulates at the higher elevations and is once more deposited in streambeds through sporadic intervals of mild temperatures.

Table 4.6

Monthly Distribution of Diversions
from Daggs Reservoir
1946

Month	Percent of Annual Water Diverted from Silver Creek
January	2
February	2
March	5
April	13
May	15
June	17
July	16
August	12
September	9
October	5
November	2
December	2
Total	100

SOURCE: Bureau of Reclamation (1947:72).

Due to the marked eccentricity of temperature and precipitation in the basin, significant deviations from the described cycle may occur during any particular year. For example, monthly discharge in the Little Colorado River at Holbrook for the years 1905-07 displayed significant variation for the same month during different years (see Table 4.8). Similarly, Lyman Dam and Reservoir, which is located 4 miles south of St. Johns and which contains a storage capacity of 20,600 acre-feet, received an average annual discharge of 18.7 cubic feet per second (approximately 13,540 acre-feet per year) for the years 1940-1957 (Akers 1964:7). Analysis of surface water discharge into Lyman Reservoir revealed that during this period actual discharge ranged from 16,000 cubic feet per second to at times no discharge at all. Furthermore, storage within the reservoir ranged from an overflowing 25,500 acre-feet in May of 1941 to several occasions when reservoir was completely dry due to the discharge being below the amount of water expended for irrigation (*ibid.*).

Because surface water flow in the region varies in direct response to precipitation, the volume and velocity of water flowing in a riverbed may change dramatically from one week--indeed, one day--to the next. Several streams, most notably the Little Colorado at lower elevations,[16] have been transformed within hours from dry sandy riverbeds to raging torrents, destroying bridges, dams and other obstacles in their path.[17] While the mean seasonal flow of water in the Little Colorado River near Holbrook is 309 cubic feet per second (Dames and Moore 1973 [section 4]:143), the U.S. Corp. of Engineers has calculated that during the peak flood of September 19, 1923 the discharge was 60,000 cubic feet per second (*ibid.*). Similarly, the maximum flow at Holbrook during the flood of 1915, when Lyman Dam above St. Johns burst,[18] is estimated to have been as high as 189,000 cubic feet per second (*ibid.*).

The responsiveness of surface-water flow to the vagaries of precipitation in the basin is also illustrated by the extreme variation in annual discharge in the Little Colorado River at specific locations in the region. While the average annual discharge of the Little Colorado at Woodruff for the period 1940-1947 was 86,000 acre-feet, during 1944 annual water flow at this location totaled only 20,000 acre-feet. During 1944, however, the volume of water discharged near Woodruff was 280,000 acre-feet, or nearly 15 times the 1944 figure, (Bureau of Reclamation 1950:7). Similarly, discharge figures for the Little Colorado River at Holbrook (a few miles downstream from Woodruff) for the years 1950-1969 show a range of variation from less than 16,000 acre-feet during 1950 to nearly 200,000 acre-feet during 1968 (see Dames and Moore 1973:Plate 4.4.3-4; see Figure 4.2). Streamflow figures at St. Johns for the years 1930-1944 exhibit the same pattern of variation.[19] The average annual discharge during this period was 8,430 acre-feet.[20] This average, however, masks a variation of between 2,790 acre-feet for 1942 and 50,010 acre-feet during 1941. Averages, thus, have little meaning where variation is both so pronounced and so pervasive.

Because most streams in the region (including the Little Colorado at lower elevations) are ephemeral and flow largely in response to rainstorms, most contain considerable quantities of sediment, especially during periods of heavy flow. Heavy sediment concentrations are particularly characteristic of streams in the northern portion of the study area. At these lower elevations with their alkaline soils, sediment loads may account for as much as 20% of stream low (Bureau of Reclamation 1950:3). Even Silver Creek, which originates from a clear mountain spring, acquires an increased sediment content as it descends towards the Little Colorado and receives water from its own tributaries along the way (see Map 4.3).

Table 4.7

Snowpack Depth on Mount Baldy
1950 - 1964
(in inches)

Year	January 15	February 1	February 15	March 1	March 15	April 1
1950	14	13	14	12	8	0
1951	11	19	12	26	12	t
1952	32	38	38	37	55	41
1953	21	19	21	23	26	15
1954	20	22	19	17	11	16
1955	--	20	17	19	10	--
1956	t	--	24	20	14	0
1957	8	23	11	9	8	0
1958	11	14	22	30	47	34
1959	5	t	11	13	t	0
1960	37	33	37	37	27	15
1961	15	22	16	17	19	15
1962	41	49	40	53	55	47
1963	18	18	26	16	18	8
1964	2	10	11	18	20	12
X	17	21	21	23	22	15

SOURCE: Enz and Weller (1964).
NOTE: Station elevation was 9,125 feet; "t" equals trace.

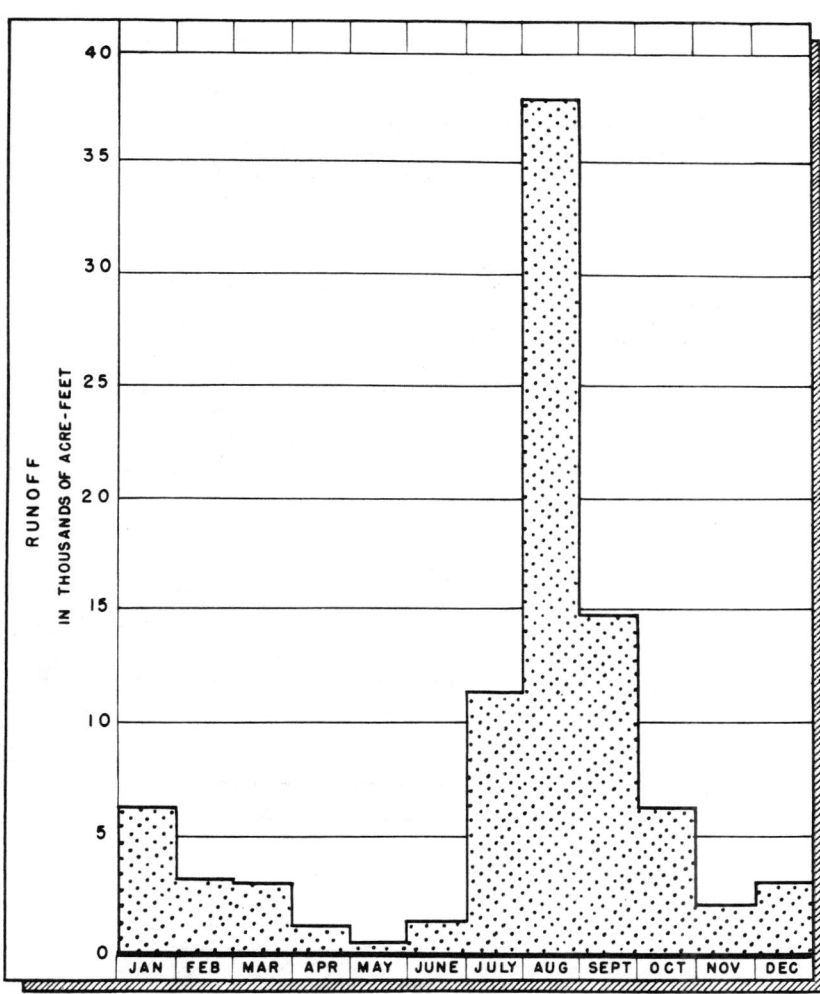

Figure 4.1
AVERAGE MONTHLY RUNOFF
LITTLE COLORADO RIVER AT HOLBROOK

Source: Dames & Moore (1973: Plate 4.4.3-5)

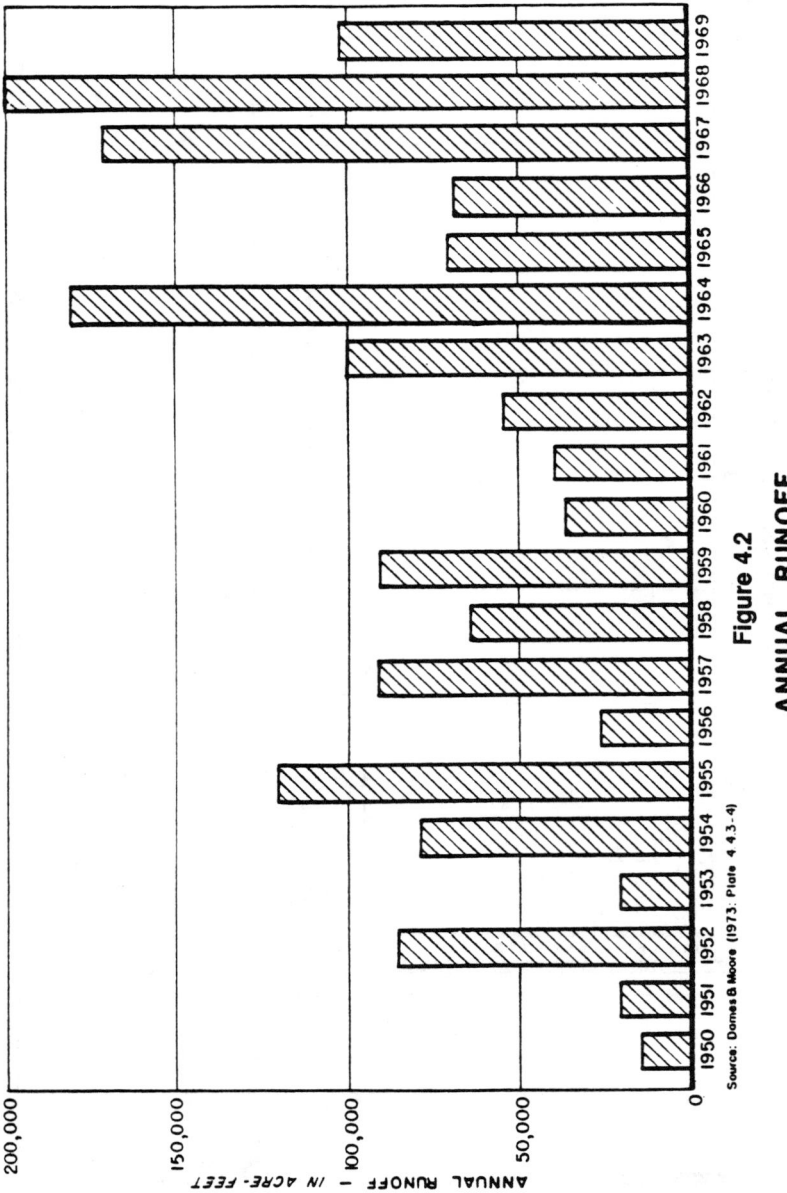

Figure 4.2
ANNUAL RUNOFF
Little Colorado River at Holbrook

The Little Colorado River Basin 99

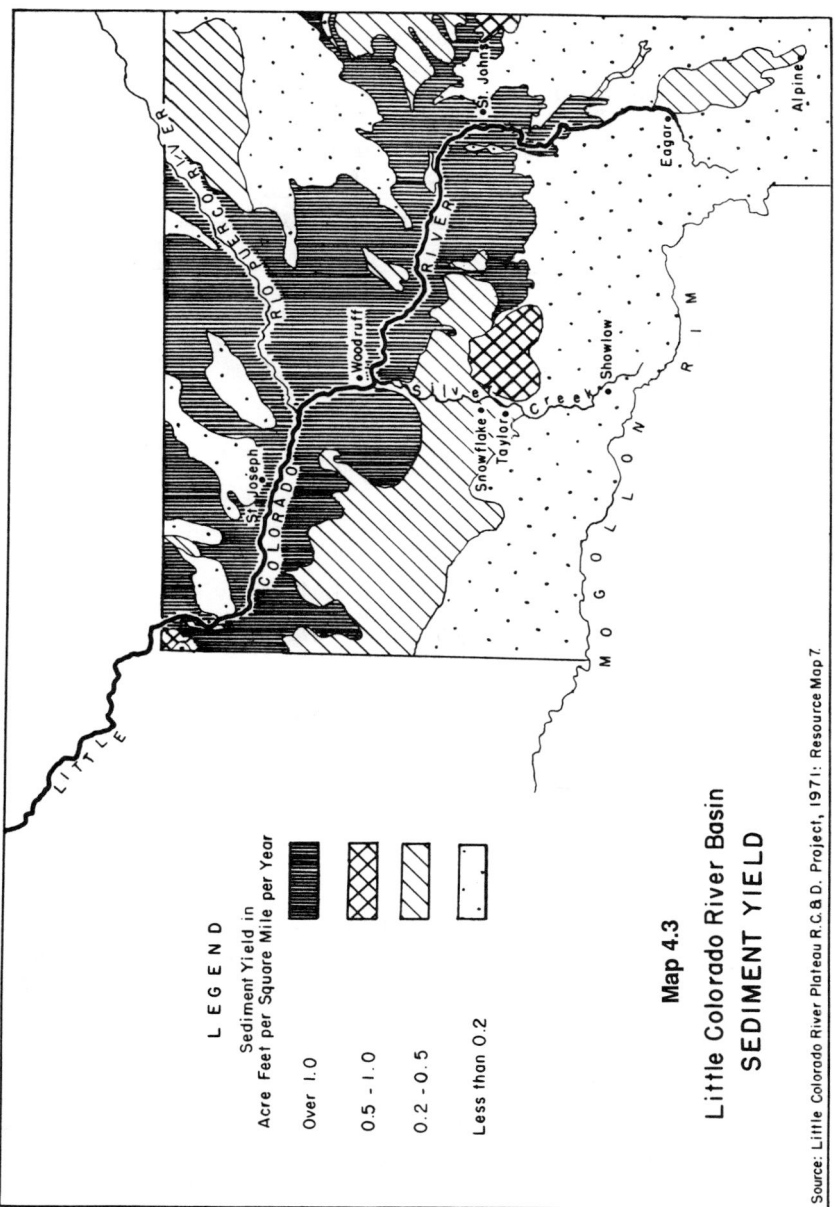

Map 4.3
Little Colorado River Basin
SEDIMENT YIELD

Source: Little Colorado River Plateau R.C.& D. Project, 1971: Resource Map 7.

Table 4.8

Monthly Discharge of the Little Colorado River
at Holbrook, Arizona
March 1905 - April 1907
(in acre-feet)

Month	1905	1906	1907
March	25,700[a]	38,200	27,300
April	54,400	14,600	23,900
May	- - -	3,320	- - -
June	4,290	244	- - -
July	4,180	1,530	- - -
August	10,000	4,400	- - -
September	18,000	4,090	- - -
October	3,120	1,640	- - -
November	69,000	672	- - -
December	6,950	11,100	- - -
January	27,800	17,000	- - -
February	9,440	9,780	- - -

SOURCE: LaRue (1916:108-109).
[a]Discharge for March 17-31 only.

Due to their intimate association with the founding and development of most early Mormon settlements in the basin and with the history of the region as a whole, two streams deserve specific consideration. These are the Little Colorado River and Silver Creek, its principal tributary in the study area.

The Little Colorado River
The Little Colorado River begins as two small mountain streams created by snowmelt originating on Mount Baldy. This water, sparkling clear and cold, flows rapidly as it descends the steep slopes of the White Mountains. At this elevation, salt and other sediment concentrations in the river are practically nil (Greenwood 1960:85). Before reaching Springerville and Eagar, the river flows through a channel which alternately widens and narrows and contains several locations

suitable for irrigated farming were the growing season of sufficient average duration.

Between Springerville and St. Johns the Little Colorado flows in a narrow, shallow channel bordered by willows, cottonwoods and saltgrasses with an average gradient of 37.6 feet per mile (Harrell and Eckel 1939:29).[21] A perennial flow exists in the Little Colorado south of St. Johns, being sustained by several large springs in this area (Akers 1964:8).

Below St. Johns the river changes dramatically. The gradient declines sharply to around 10 feet per mile (*ibid.*:28) as the Little Colorado changes direction and begins its northwesterly flow, bisecting the basin. At this point, the river opens into a flat and sandy channel. Surface flow becomes intermittent, with most of the flow during the late spring being confined to subsurface channels. The Little Colorado retains this appearance until well after it leaves the region. The river channel near Holbrook and Joseph City has been described as

> broad and alluviated. Typical of such streams, it continually abrades its channel and redeposits the transported material downstream. As a result, the channel conditions change from time to time, such that there is no stable relationship between the stage of the river and its discharge. At present, the channel is over 100 feet wide...and is 5 to 10 feet deep. The banks are lined with riparian plants, such as salt cedar and mesquite. (Dames and Moore 1973 [section 4]:41)

Water quality in the Little Colorado south of St. Johns is quite good. However, as this water flows downstream, traversing high salt-bearing formations and acquires additional water from numerous tributaries originating in these saline strata, its purity is quickly lost. Water quality in the Little Colorado River is of questionable value below St. Johns, where the concentration of both suspended and dissolved solids (particularly soluble salts) renders it largely useless for domestic, agricultural or even industrial use (see Tables 4.9[22] and 4.10).

Located some 15 miles downstream from St. Johns are Zion Dam and Reservoir. Originally built between 1902 and 1905, Zion Dam was reconstructed in 1908 with a reservoir capacity of 12,896 acre-feet and a surface area of 2,048 acres (Akers 1964:8-9). By 1952, 22,700 acre-feet of sediment had been deposited behind the dam, reducing its storage capacity to 760 acre-feet (*ibid.*). Continued silting has terminated the use of this reservoir.[23] Sediment buildup has even significantly reduced the storage capacity of Lyman Reservoir, 5 miles south of St. Johns.[24]

Stream flow variability in the Little Colorado becomes more pronounced downstream, as the quality of its water deteriorates, because the river drains an increasingly larger area, and a greater proportion of its flow derives from tributaries originating to the north. Most of the water emanating from these less desirable northern tributaries enters the Little Colorado during intense summer storms and, thus, predominates in the stream at the height of the irrigation season.

These waters almost always contain excessive amounts of sodium salts and other sediment, with a large portion being particularly fine-textured silt derived from shale (Bureau of Reclamation 1950:9).

Table 4.9

Suspended Sediment Concentration
in the Little Colorado River
at Cameron, Arizona
1969

Date of Sample	Suspended Solids Concentration (mg/l)[a]
January 30	27,800
March 27	19,700
July 22	99,500
August 26	56,800
X	51,000

SOURCE: Dames and Moore (1973 [Section 4]:151).
[a]Milligrams per liter.

The sediment load carried in the Little Colorado River near Holbrook and Joseph City is always high, prompting its description as "red, turbid and unaesthetic" (Dames and Moore 1973 [section 4]:149). Dissolved solids normally exceed 500 mg/l during average discharge conditions. During periods of high evaporation, however, dissolved solids increase to values greater than 1,000 mg/l (*ibid.*:149). Such water is only suitable for irrigating salt-tolerant crops in well-drained soils.

Silver Creek
While not necessarily providing perennial abundance, Silver Creek, in marked contrast to the Little Colorado River and other local streams, presents at least a picture of persistent moderation. It does not portray the regionally-distinctive extremes of excess and privation. In addition, chemical analyses have consistently determined the waters in Silver Creek to be relatively pure and free from the heavy concentration of salt and other solids which plague the Little Colorado and its other tributaries.

Table 4.10

Chemical Concentrations at Selected Sites
along the Lower Valley of the Little Colorado River

Site Location	Total Solids Concentration (mg/l)	Sodium (Na)[a] Concentration (mg/l)
Holbrook	868	185
Penzance 1	507	98
Penzance 2	682	130
Havre 1	2,340	728
Havre 2	877	200
Winslow 1	3,060	984
Winslow 2	2,340	719
Winslow 3	2,030	610
Winslow 4	2,080	636

SOURCE: Dames and Moore (1973[section 4]:148).
NOTE: Site locations are as follows:

Penzance 1 and 2 are two miles west of Holbrook.
Havre 1 and 2 are twelve miles west of Joseph City.
Winslow 1 and and 2 are four miles east of Winslow.
Winslow 3 is two miles east of Winslow.
Winslow 4 is five miles north of Winslow.

[a]Figures for sodium concentration also include Potassium (K). However, a chemical analysis of Little Colorado River water near Woodruff shows the concentration of potassium to be quite small, 3.48 mg/l on the average.

Originating from a spring located about 10 miles southeast of Snowflake and Taylor, Silver Creek comprises the only fully-perennial stream in the Little Colorado River Basin. Silver Creek Spring, which actually discharges water from several neighboring locations, is the largest spring in the entire region, with a rate of flow estimated at 11 second-feet (cubic feet per second; Harrell and Eckel 1939:67) and ranging between 8 and 13 second-feet (Bureau of Reclamation

1947:13). The average annual discharge in Silver Creek downstream from Snowflake for the 21 year period between October, 1950 and September, 1971 was 11,600 acre-feet (Salt River Project 1974 [section 3]:283).

More significant than its annual discharge is the stability displayed by Silver Creek Spring. Harrell and Eckel (1939:30) reported an inability to locate even one account of failure in the flow of water from this spring, "even in the most severe drought." A Bureau of Reclamation (1947:50) study of the Snowflake area corroborates the finding of Harrell and Eckel, stating that "local information covering a period of nearly 50 years indicates that there is little variation in the flow" of Silver Creek Spring. This same report (*ibid.*:87) alludes to a sworn statement dated August 25, 1904 (at the end of a six-year drought in the basin) that the flow of water in Silver Creek was 13.5 second-feet, and concludes that

> although lower flows of Silver Creek have been recorded in recent years, it is evident that the springs feeding it provide a base flow relatively unchanged by drought.

One important consequence of the stability of Silver Creek Spring from the perspective of establishing viable agricultural communities has been the reliability of available water in the stream that it produces. While Silver Creek, like all other streams, is responsive to variations in snowmelt and precipitation, an ostensible lower limit of water availability exists in its channel, making Silver Creek a substantially more dependable source of surface water than any other stream in the Little Colorado River Basin (see Table 4.11). The degree of variation displayed by the flow of Silver Creek into Daggs Reservoir differs sharply from that of the Little Colorado River either at St. Johns or in the Lower valley, in both its monthly and annual discharge figures. While the relative stability of annual discharge in Silver Creek has provided sufficient predictability upon which to base agricultural activities from one year to the next, the stability associated with its monthly discharge has resulted in a reliable flow of water during the critical dry months of April, May and June. No report exists which indicates that Daggs Reservoir has ever been completely dry due to insufficient flow in Silver Creek.

Downstream, near Woodruff, the picture of Silver Creek changes. Because much of Silver Creek's water is expropriated by Snowflake, Taylor and Shumway,[25] and because its drainage area increases markedly at lower elevations, annual and monthly differences in discharge become more pronounced, approaching the fluctuations characteristic of the Little Colorado River (see Tables 4.12, 4.13 and 4.14). While annual discharge in Silver Creek for the years 1942-1944 averaged 13,000 acre-feet at Daggs Reservoir, runoff near Woodruff averaged only about 8,100 acre-feet. On the other hand, the annual discharge in Silver Creek near Woodruff in 1941 was over 35,000 acre-feet, and in 1932 the discharge exceeded 59,000 acre-feet. Such increased intensity of water flow, rarely experienced near the head-waters of Silver Creek, places considerable stress upon the physical structures underlying irrigated agriculture and adds significantly to the cost of maintaining agricultural systems.

Table 4.11

Monthly Discharge of Silver Creek into Daggs Reservoir
1942 - 1944
(in acre-feet)

Month	1942	1943	1944	$\bar{X}$	Percent of Total
January	713	763	683	713	6
February	1,194	1,734	654	1,194	11
March	2,024	2,762	3,528	2,771	25
April	961	531	1,017	837	8
May	788	533	654	659	6
June	750	513	642	635	6
July	729	522	691	647	6
August	931	521	708	720	6
September	802	585	670	686	6
October	816	644	704	721	7
November	791	662	676	710	6
December	801	738	652	730	7
Total	11,300	10,508	17,279	11,023	100

SOURCE: Bureau of Reclamation (1947:54).

Dam That River!

Table 4.12
Streamflow Summary for Silver Creek near Woodruff
1929 - 1946
(in acre-feet)

Water Year	October	November	December	January	February	March	April	May	June	July	August	September	Total
1929-30	698	484	323	508	1,710	5,830	676	56	42	3,930	6,980	440	21,677
1930-31	284	629	740	940	3,930	1,250	393	377	292	3,060	3,850	3,240	18,985
1931-32	986	1,090	1,220	1,370	37,800	12,700	613	248	61	2,230	912	250	59,480
1932-33	1,110	825	730	732	1,320	5,380	42	181	507	5,380	3,180	4,470	23,857
1933-34	---	---	---	---	---	---	---	---	---	---	---	---	---
1934-35	---	---	---	---	---	---	---	---	---	---	---	---	---
1935-36	1,030	639	795	760	2,280	1,010	1,020	93	126	3,490	3,200	1,500	15,943
1936-37	623	484	371	448	9,730	9,400	294	151	272	511	477	282	22,683
1937-38	85	101	151	103	76	2,750	117	0	4	143	2,840	1,190	7,560
1838-39	52	75	79	331	383	1,670	807	4	0	141	1,680	462	9,884
1939-40	204	375	226	428	476	175	117	0	91	6,850	3,010	3,690	15,642
1940-41	895	349	1,780	3,850	3,520	16,170	3,810	165	180	841	1,650	2,060	35,270
1941-42	1,890	153	605	1,130	307	791	325	114	39	4	372	55	5,785
1942-43	456	254	448	325	375	1,310	294	54	15	798	3,720	1,000	9,049
1943-44	472	189	456	670	662	4,460	835	112	33	542	679	501	9,591
1944-45	188	226	357	363	426	3,370	93	1	12	2,270	5,280	160	12,746
1945-46	480	177	309	417	341	189	3	0	0	2,200	4,220	6,140	14,456
X̄	629	402	573	825	4,125	4,430	629	100	112	2,159	2,803	1,696	18,840
%	3	2	3	4	22	24	3	1	1	11	15	9	98*

Source: Stubblefield (1953:22).
*Total does not equal 100% due to rounding error.

Table 4.13

Annual Discharge at Selected Locations
in the Little Colorado River Basin
1930 - 1944
(in acre-feet)

Year	Little Colorado River at St. Johns	Little Colorado River at Woodruff	Silver Creek at Woodruff	Silver Creek at Snowflake[a]
1930	5,280	42,700	21,700	7,700
1931	3,120	64,200	19,000	9,300
1932	15,600	117,000	59,500	16,200
1933	5,050	51,600	23,900	7,400
1934	- - -	- - -	- - -	8,000
1935	- - -	- - -	- - -	9,600
1936	4,350	42,710	15,940	8,700
1937	3,540	46,100	23,040	19,600
1938	4,110	15,110	7,560	8,800
1939	2,460	10,370	5,680	7,900
1940	9,750	46,100	23,040	10,100
1941	50,010	115,400	35,270	19,900
1942	2,790	14,560	5,790	11,300
1943	2,080	20,250	9,050	10,500
1944	1,460	14,850	9,590	11,300
X	8,430	46,230	19,360	11,080

SOURCE: Bureau of Reclamation (1947:58); Colorado River Commission of Arizona (1940[section 4]:1).
NOTE: Figures are rounded to nearest tenths.
[a]Figures for Snowflake comprise runoff into Daggs Reservoir, and those from 1930 to 1942 are estimates of actual runoff at this location. All figures for Snowflake are rounded to nearest hundreds.

Besides providing a reliable quantity of water, Silver Creek differs significantly from the Little Colorado and its other tributaries by furnishing excellent quality water most of the time as well. Except for the periods of peak flow associated with heavy rains, the concentration of suspended solids is generally below 25 parts per million, and even during periods of heavy flow this figure is normally exceeded only in the lower stream near Woodruff (see Table 4.15). A Bureau of Reclamation (1947:66-67) analysis of the chemical quality of Silver Creek water concluded:

> Results of analyses of Silver Creek waters, sampled near Woodruff, indicate the quality to be suitable for irrigation use. Complete discharge data for the stream at the point sampled were not available, so weighted averages of the concentration of waters could not be determined. The highest concentration of waters sampled, however, was but 440 parts per million, and the percent sodium exceeded 38 in only one instance, indicating the water to be of good quality. Water stored in Daggs Reservoir and that in Silver Creek above the reservoir is of excellent quality. The water in the reservoir at the close of the 1942 irrigation season had a concentration of 160 parts per million and a sodium percentage of 12, while that in Silver Creek above the reservoir has an average concentration of 119 parts per million and an average sodium percentage of 25.[26]

As the report itself points out (*ibid.*), the best proof of the superior quality of Silver Creek lies in the fact that, (in contrast to Joseph City), continued irrigation with Silver Creek water for nearly 70 years had produced no noticeable soil deterioration in the Snowflake-Taylor area. Thus, in both the quantity and the quality of its water, Silver Creek offers an abundant and reliable resource base upon which to establish an agricultural community.

HUMAN IMPACT ON THE LITTLE COLORADO RIVER BASIN

It would be incorrect to assume that current environmental conditions in the Little Colorado River Basin represent those encountered by the early pioneers settling this region. Due to the fragility of this largely arid environment, important changes have resulted from the impact of human resource exploitation. Two major impacts will be discussed--overgrazing and irrigation--in order to more fully appreciate the condition of the natural environment prior to Mormon immigration. The discussion that follows bears directly on the issue of differential community development during the settlement period.

Overgrazing

Several years ago, Colton (1937) published an account which clearly illustrated the devastating impact of overgrazing on one location within the Little Colorado River Basin.[27] While excavating a house which he believed to be a prehistoric pit house--the floor of the house was 30 inches below the existing level of the river bank--Colton discovered that, in fact, the house had been built in either 1878 or 1879. Through subsequent interviews, he further discovered that the house in question was still in perfect condition as late at 1884. At that time, the flat lands on either side of the river supported a fine stand of both old and young cottonwoods, while gama grass covered most of the surrounding hills. The house then stood about 100 feet from the river, and beaver were known to inhabit the stream, living off the cottonwood trees. However, in 1884 several thousand head of sheep were imported into the valley. These sheep were apparently maintained without noticeable range deterioration until the drought of the early 1890's. At this time, Navajos reportedly entered the area to cut down the cottonwoods and feed them to their starving herds. When the rains resumed in the mid-1890's with no grass to hold the water, disastrous floods ensued. Because the riverbed was narrow at this location and could not carry much flood water, the river overflowed its banks and deposited 30 inches of silt by 1935. When Colton returned to the excavation site in 1937, the river had widened another 14 feet and only the back wall of the house was still standing.

Table 4.14

Variation in Monthly Discharge in Silver Creek
1942 - 1944

	Daggs Reservoir	Woodruff
Months included	36	36
Mean monthly discharge	919.1	623.6
Standard Deviation	633.661	915.028
Coefficient of Variation	.689	1.461

SOURCES: Tables 4.11 and 4.12.

Widespread grassland deterioration has occurred throughout the region as a result of the persistent overstocking of ranges.[28] Cattle and sheep numbers on ranges in the basin were quite high between 1885 and 1925. Apache County assessment rolls registered an average of 35,119 head of cattle and 117,762 head of sheep for the years 1916 through 1925, compared to 19,630 cattle and 3,882 sheep for the years 1958 through 1967 (see Table 4.16).[29]

Recovering from the Depression of 1873, the cattle industry throughout Arizona expanded rapidly. "Ranges that for permanent and regular use would have been overstocked with one cow to every 100 acres were overloaded until they carried one cow to every 10 acres" (Spencer 1966:21). Range deterioration within the Little Colorado River Basin began in the 1880's when the Aztec Land and Cattle Company purchased over 1 million acres and imported more than 60,000

Table 4.15

Water Quality in Silver Creek at Woodruff and at Daggs Reservoir
(in parts per million)

Date of Sample	Woodruff		Daggs Reservoir	
	Total Dissolved Solids	Sodium	Total Dissolved Solids	Sodium
June 23, 1942	374	31	---	10
September 30, 1942	340	23	160	5
November 9, 1942	na	na	116	11
November 25, 1942	316	17	na	na
April 27, 1943	324	29	na	na
May 1, 1943	na	na	122	10

SOURCE: Bureau of Reclamation (1947:68, Table 17).

head of cattle onto this range (see Chapter 7). While the extended drought of the early 1890's forced the Aztec Company into bankruptcy, the overstocking of ranges had already radically altered existing range conditions.[30]

Overstocking of ranges is still a problem in the region. Due to the unpredictable nature of precipitation throughout the region, ranchers experience great difficulty in adjusting cattle numbers to changing moisture conditions. They, therefore, stock their ranges for conditions intermediate between high and low extremes (Salt River Project 1974 [section 3]:42). Given the extensive overstocking that occurred previously, this prevailing management strategy (see also Underwood 1970) has preserved the deteriorated condition of the range. Consequently,

> the once almost pure stands of winter fat on the heavy alkaline-free soils have been virtually eliminated and replaced by snakeweed and rabbitbush. The broad expanse of grass that once covered most of Navajo and Apache Counties has deteriorated under heavy grazing pressure and become overrun with snakeweed and pinque. (Salt River Project 1974 section 3:76)

Several indicators signify the existence of overstocked ranges. One is the relative predominance of non-palatable perennial grasses. As already indicated, the herbaceous class of vegetation accounts for only a small fraction of either the grassland or woodland communities. Palatable shrub species are similarly scarce, and the frequency of unpalatable grasses such as ring muhly is high. Galleta, the most commonly occurring grass, is practically useless for grazing when dry --indeed, it can be lethal if consumed by livestock in sufficient quantities while in a dry state (see Schmutz 1968).

Another indication of overgrazing is the expansion of juniper trees into the grassland community. The grazing habits of livestock provide the conditions which facilitate juniper expansion (see Nichol 1937:191). Juniper seeds germinate only after they have passed through the alimentary tract of an animal. Livestock, thus, facilitate the dissemination of juniper seeds at the same time that they remove the competing grasses which inhibit juniper expansion. Controlling juniper expansion has been a major concern of ranchers in recent years. As one drives though that portion of the region in which juniper predominate (most notably near Snowflake), extensive stands of juniper may be seen uprooted as a result of widespread bulldozing--"chaining" as it is called--in an effort to control their spread and to increase the number of palatable grasses available for cattle.[31]

Extensive range deterioration has significantly reduced range productivity. Rangeland in the Joseph City area is producing 225 pounds of forage per acre per year, which is only 45% of the 500 pounds of forage per acre per year estimated as the potential for this location (Dames and Moore 1973 [section 4]:225). Consequently, only between 6 and 7 animal units can be supported per section in the Joseph City area (*ibid*.:235).[32] Estimates of the number of animal units supportable on a section of rangeland in the Snowflake and St. Johns areas are 3.5

and 9.9 respectively (Salt River Project 1974 section 3:85, 105). These figures represent, according to local authorities, perhaps one-third to one-half of the number of animal units supportable prior to the 1880's.[33] A 40% reduction in current grazing level has been suggested as the requirement necessary to restore the ranges to their productive potential (Dames and Moore 1973 [section 4]:225).

Table 4.16

Number of Cattle and Sheep Registered
in Apache County
1916-1925 and 1958-1967

Year	Cattle	Sheep	Year	Cattle	Sheep
1916	25,093	107,197	1958	22,472	5,202
1917	30,701	143,961	1959	20,939	5,129
1918	37,778	144,177	1960	21,291	4,716
1919	33,550	145,846	1961	20,241	5,121
1920	38,362	132,494	1962	19,731	5,733
1921	35,351	113,108	1963	19,083	2,530
1922	38,413	120,432	1964	18,489	2,769
1923	41,386	114,772	1965	16,688	2,425
1924	38,440	80,603	1966	18,488	2,447
1925	32,114	75,031	1967	18,881	2,747
X	35,119	117,762	X	19,630	3,882

SOURCE: Apache County (n.d.).

Besides reducing livestock productivity, over-grazing has had a deleterious impact on surface water quality in the basin and ultimately on the production of those crops which depend on surface water for irrigation. The reduction of biomass caused by excessive grazing pressure (particularly the high proportion of bare surface produced by overgrazing) has enhanced soil erosion and added considerably to the silt-bearing quality of the Little Colorado and other streams in the basin. One U.S. Department of the Interior report (1946:152) suggests that in one year the Little Colorado River transports "the equivalent of nine inches of topsoil from an entire township." Overgrazing within the region, as well as that which has occurred on the Navajo Reservation to the north, has contributed substantially to the deterioration of the Little Colorado River at lower elevations and of farming operations which employ this water source (see Abruzzi 1985).

Irrigation

Although continuous irrigation has not caused any appreciable soil deterioration among settlements at higher elevations, the same cannot be said for towns situated in the lower valley. Joseph City has suffered the greatest dependence on the poorest quality soils and surface water of any Mormon settlement in the region. Marked differences in soil and surface-water quality existed prior to the advent of farming and ranching within the basin. However, these differences increased as the water available to farmers in the lower valley deteriorated following Mormon settlement of the region in the 1870's. Other settlements were established upstream which eventually acquired prior use of waters flowing in both the Little Colorado River and Silver Creek. Thirty direct irrigation rights exist on the upper Little Colorado River which have been amalgamated into two irrigation companies: The Round Valley Water User's Association and the St. Johns Irrigation Company. Many separate irrigation rights have also been incorporated into irrigation companies serving the towns of Snowflake, Taylor, Shumway, Showlow and Woodruff along Silver Creek. In addition, several small, private water rights exist on other tributaries to the Little Colorado. By 1930, when all existing storage reservoirs in the Little Colorado River Basin had already been completed, 37 reservoirs impounding 72,795 acre-feet of water had been constructed (Bureau of Census 1930).

As a result of the prior appropriation of water, settlers at St. Joseph had to cope with a greater proportion of their irrigation water originating in the Rio Puerco and other northern tributaries to the Little Colorado. Because this water has particularly high concentrations of salts and other dissolved solids, its continued use in irrigation has significantly reduced the productivity of soils in the vicinity of this town. Through capillary action, continued irrigation has also increased the surface accumulation of salts already present in these alkaline soils. It is not an uncommon sight to see fields in the Joseph City area blanketed with a white layer of fine silt particles which have coated the surface of the soil and baked dry in the hot summer sun. Due to its exceptionally fine texture, this silt produces a clay-like layer as it dries which is highly impervious to water. The result, in addition to salt stress on crops, has been both poor drainage and inadequate soil aeration, which further inhibit agricultural productivity.

CONCLUSION

During their colonization of the Little Colorado River Basin, early Mormon pioneers occupied several distinct and widely separated local habitats. Due to the marked variability associated with key features of the natural environment in this basin, each habitat presented a unique set of material conditions to which specific local populations had to adapt. The precise combination of subsidies and drains present within each habitat ultimately determined the return upon labor invested there and, thus, the success of local agricultural activities undertaken. Exploiting whatever subsidies were available, the early settlers frequently had to bear the costs of heavy drains upon their meager resources due to a frequently demanding and highly variable natural environment. While certain constraints proved surmountable, others remained beyond the control or the adaptive capacity of these small populations with their limited resources. Already containing important differences with respect to those conditions influencing agricultural productivity, local habitats diverged further in their agricultural potential throughout the settlement period. The key to understanding the differential development displayed by these early settlements, thus, lies in comprehending the configuration of environmental diversity within the basin--indigenous and derived--and the distinct adaptive demands which this placed on the maintenance of specific local populations. Significantly, similar developmental patterns emerged within broad sub-regions of the basin.

Because climatic conditions remained beyond human control, these conditions ultimately circumscribed agricultural productivity in the region. The existence of an arid climate throughout most of the basin absolutely precluded the viability of substantial farming operations existing outside of a few river valleys, and the numerous, abandoned settlements dotting the basin testify to the costs incurred by individuals flaunting this imperative.[34] Similarly, the inverse relation between altitude and length of the growing season and the greater variability of the length of the growing season at higher elevations combined to severely constrain agricultural productivity among settlements located in the southern highlands. Alpine, and to lesser extent Showlow,[35] consistently displayed among the smallest and most variable population size and annual productivity (as reflected in tithing collected) of any Mormon settlement in the basin during the nineteenth century.[36]

Although aridity and an inadequate growing season decisively restricted the number of suitable locations for establishing viable agricultural settlements in the basin, other considerations, somewhat less categorical in their impact, determined the viability of farming settlements at remaining locations. For those settlements beyond the pale of immediate climatic constraints, the availability of suitable surface water for irrigation was the most pressing limiting factor governing agricultural productivity. For most settlements in the basin, extreme variation in surface-water flow posed the dual problem of exceeding both the minimum and maximum tolerance limits endurable by early Mormon agricultural systems. While an insufficient supply of silt-free water precluded healthy plant growth, an excessive flow frequently placed an unbearable stress on the frail irrigation structures erected by the early pioneers, precipitating their collapse. For several reasons, however, the drains imposed by fluctuations in surface water flow weighed most heavily on settlements in the lower valley of the Little Colorado River.

Unlike towns at higher elevations to the south which were able to construct storage reservoirs and, thus, offset seasonal shortages in surface water availability, St. Joseph, Woodruff and the other early settlements in the lower valley were relegated to the use of diversion dams. Consequently, despite a heavy investment in irrigation systems, the lower valley settlements remained subject to the pervasive instability imposed by the highly variable and frequently insufficient flow of water in the Little Colorado River.

The physical stresses imposed by heavy stream flow were also most clearly experienced by the lower valley irrigation systems. Because the drainage area near St. Joseph and Woodruff is considerably larger than that near any other settlement in the basin, the absolute magnitude of water impacting the dams of these two towns during periods of heavy stream flow was considerably greater than that sustained by similar structures elsewhere in the region. At the same time, the deep, alluvial composition of the lower Little Colorado River bed aggravated the difficulties attending dam construction and increased the vulnerability of these dams.[37]

The problems of excess and privation which characterized the flow of water past St. Joseph and Woodruff were soon exacerbated by flooding upstream on both the Little Colorado River and Silver Creek. The prior appropriation of streamflow by these later settlements amplified the shortage of suitable irrigation water available to the lower valley settlements during periods of low flow.[38] Conversely, during periods of particularly heavy flow, dam failures upstream increased the destructive force of an already swollen river, frequently destroying irrigation systems and inundating fields throughout the lower valley.[39] Drought and flood conditions frequently added to the stress imposed on crops and soils in the lower valley--for the reasons already stated, but also due to the higher silt burden which both situations provided.

The lower valley settlements also suffered from lower quality soils than those found elsewhere in the basin. Lower in quality initially, the suitability of these soils for agricultural purposes declined further as a result of continued irrigation with water reduced in purity by overgrazing and by the prior appropriation of clear water upstream.

Although settlers occupying the river valleys at intermediate elevations experienced constraints upon their productivity as well, the restrictions which they encountered were not nearly as extreme as those imposed upon populations located at either higher or lower elevations in the basin. Possessing a sufficient and reliable growing season, each of the remaining settlements--Snowflake, Taylor, St. Johns and Eagar--benefitted from a dependable, and silt-free stream flow, suitable locations for storage reservoirs, a relatively mild seasonal climatic shift and about 4 inches of precipitation more than that available in the lower valley. At the same time, all of these settlements are located in broad valleys containing relatively fertile soils. Significantly, these are the same four towns described in Chapter 2 as ranking highest regarding the size and stability of their populations and agricultural productivities, and in the complexity of their community organization.

As already indicated, a principal factor contributing to differences in productivity and population size among the agricultural settlements in this arid river basin during the nineteenth century was the fluctuation that occurred in the availability of water for irrigation purposes. Integrally related to this variation in water supply--both as cause and effect--were the systemically important dam failures, which in some settlements occurred with frustrating regularity. Irrigation systems comprised the principal infrastructural investment sustaining agricultural productivity in the basin. Consequently, a dam failure imposed a severe drain on a settlement's resources and jeopardized its very survival. For this reason, the following chapter will examine the history of dam constructions and failures in greater detail.

NOTES

1. The term, Four Corners, refers to that general region surrounding the intersection of the boundaries of Arizona, New Mexico, Utah and Colorado.
2. The region under consideration actually comprises only the southern portion of the Little Colorado River Basin. The northern part of the basin consists principally of badlands located within the Navajo reservation. This latter region will be considered only as it impinges on populations inhabiting the Little Colorado River Valley along the northern periphery of the study area. For convenience sake, however, the term Little Colorado River Basin will be used to denote the region investigated.
3. The following discussion of the meteorological conditions underlying precipitation in the Little Colorado River Basin derives from Jurwitz (1954). The purpose for presenting this discussion is to underscore the unpredictability of precipitation and, therefore, the potential for instability that precipitation poses for agricultural communities in this basin.
4. So arid is the environment in the basin during late spring and early summer that as the initial storms enter the region, much of the precipitation is reabsorbed back into the clouds through evaporation prior to its reaching the ground. Also, many of the summer storms, particularly those that occur early in the season, represent localized cloud systems which may be viewed depositing enormous amounts of water along specific paths while locations immediately outside a cloud's line of travel remain completely untouched.
5. One hundred and twenty days represents the average growing season needed to support most crops.
6. Only fragmentary data exist on the length of the growing season at Showlow, not enough to warrant inclusion in Table 4.3. The information which does exist suggests that the growing season at Showlow is considerably less than that at Snowflake and possibly less than 120 days on the average.
7. See the section below on overgrazing for a discussion of permissible animal densities.
8. Figures on the proportion of piñon pine at different locations in the juniper-piñon woodland community are not available, but variations in their relative abundance are visibly noticeable. Because the region under investigation forms an undulating plain, rising

with an increasing gradient, changes in vegetation cover occur over shorter distances as one proceeds southward through the area. Furthermore, since vegetation within an arid climate is particularly responsive to minor shifts in precipitation, which in this area is a function of altitude, visible differences occur in plant density within localized segments of the juniper-piñon community. Even slight variations in elevation associated with the undulating character of the topography produce noticeable, localized differences in plant density between contiguous higher and lower sections of the rolling landscape. The 200% increase in plant density which occurs between St. Johns and Snowflake is associated with less than a 1 1/2 inch increase in mean annual precipitation.

9. National forests were established early in this century and eventually included a significant portion of the montane forest community. The Forest Service currently controls about 962,000 acres within the basin, most of which is located above 6,000' elevation. The Forest Service alone controls about 20% of all the land within the region, while total government administration regulates nearly 50% of the land in the basin (see Little Colorado River Plateau Resource Conservation and Development Project 1971:12).

10. The remaining land area consists largely of isolated riparian communities, such as those within the narrow river valley associated with Silver Creek. Such communities display little areal extent in relation to that of the total region.

11. The formation of the Holbrook Anticline (northwest of Snowflake) resulted in extensive faulting throughout the basin. Although several towns in the region, including St. Johns (see Salt River Project 1974 (section 3):291), have encountered problems of water storage in reservoirs due to faulting, the greatest difficulty has been experienced by Snowflake and Taylor which are located closest to the anticline formation (see Chapter 5, footnote 34).

12. Although of primary developmental importance to the communities in this basin during more recent years, groundwater played a relatively insignificant role in the adaptation of early populations. While shallow wells were dug for domestic purposes, extensive use of sub-surface water for irrigation did not begin until the 1920's (at Joseph City; see the *Snowflake Herald*:3/21/1924) and did not become widespread until after World War II. The most intensive exploitation of groundwater sources has taken place since 1960 with the introduction of large industrial enterprises into the basin. For this reason, a discussion of the variations in quantity and quality of sub-surface water will not be included here. See Abruzzi (1985) for a discussion of the relationship between community development and the exploitation of groundwater resources.

13. Greenwood (1960:4) refers to estimates which claim that over 98% of all precipitation in the Little Colorado River Basin is "used by plants, evaporated, or lost to underground tables".

14. While a sufficient supply of water generally exists in the Little Colorado River near Joseph City prior to and at the very beginning of the irrigation season, one Bureau of Reclamation study reported that from April to June, stream flow in this area is "usually inadequate to meet irrigation requirements, with no flow at all in June during 6 of the 8 years of the period studied" (Bureau of Reclamation 1950:7).

15. It was precisely those freshets associated with the advent of summer storms which wreaked so much havoc on the dams built by early pioneers in the lower valley of the Little Colorado River (see Chapter 5).

16. The intensity of variation in water flow increases as one proceeds downstream in the Little Colorado River, since the size of the drainage area increases with greater distance

118   Dam That River!

from the river's point of origin on Mount Baldy. While the Little Colorado drains only 747 square miles near St. Johns, the drainage area in the vicinity of Holbrook and Joseph City increases dramatically to 17,600 square miles.

17. This capacity of the Little Colorado and other streams in the basin to rapidly transform themselves is illustrated by the predicament of the settlers throughout the lower valley during 1876 and at Woodruff during 1878 (see Chapter 2). In these two instances, crops were lost to both drought and flooding during the same agricultural season. Such occurrences, which were not uncommon, underscore the intensity and unpredictability of environmental variation within the basin.

18. See Chapter 5, footnote 28.

19. Because the different gauging stations are part of the same river basin, the pattern of variation in water flow is correlated from one station to the next. Thus, the discussion presented here, illustrating annual variations in water flow for distinct locations during different time periods, underscores the pervasiveness of surface water variability throughout the region.

20. The Salt River Project (1974 [section 3]:290) indicates that the average discharge of the Little Colorado River above Lyman Dam over a 32-year period was 14,420 acre-feet.

21. The river channel at this point is less than 50' wide and less than 5' deep (Akers 1964:8).

22. Cameron is actually located over 70 miles downstream from the study area and much further to the north. At this location, the Little Colorado River receives most of its water from the heavily sedimented tributaries originating in the badlands north of the study area. The figures for Cameron, therefore, are not representative of the region between St. Johns and Joseph City; rather, they are presented to illustrate the extent of sediment concentration typical of tributaries such as the Rio Puerco, which enter the Little Colorado from the north.

23. One figure quoted for the annual silt discharge in the Little Colorado River is 27,500 acre-feet (Porter n.d.b.:8). Early pioneers complained often of the poor quality of water obtained from the Little Colorado, both for domestic use and for irrigation purposes. One story relating to the original pioneers in 1876 and retold frequently is particularly revealing.

> A 7-gallon kettle was filled when they camped for the night with water from this stream and set by for use the next morning after it had "settled" --there was about an inch at the top of the kettle of fairly "clear" water; but soluble matter in the water was still so much in solution that the water was of poor quality. (Porter, n.d.b.:7-8)

24. Leone (1979:92-93) discusses the problem of silt accumulation in reservoirs along the Little Colorado and argues that dam washouts in the 19th century were essential to successful farming in this region. He Maintains that once permanent dams were constructed in the 1920's, they accumulated silt and became useless--even dangerous. Due to silt accumulation and the consequent raising of the river bed for miles behind a dam, water was more likely to spill over the dam and flood surrounding fields. Holbrook, which is located 10 miles above the Joseph City dam at Penzance, is continuously threatened with flooding as a result of silt aggradation behind that dam.

25. Shumway is a small town of approximately 100 persons located a few miles south of Taylor.
26. As measures of chemical concentration in water, parts per million and milligrams per liter may be considered equivalent below concentrations of 5,000 ppm.
27. While the situation described by Colton occurred at a location several miles downstream from the town of Winslow, and thus outside the study area, the forces which contributed to the massive erosion that he discovered have been ubiquitous throughout the Southwest, including that portion of the Little Colorado River Basin under consideration here.
28. Range deterioration has been greatest in the grassland and juniper-piñon woodland communities.
29. If the figures for the years 1924 and 1925 are discounted, the average number of sheep during the earlier period becomes 127,748 head. The sheep industry has completely disappeared from the study area (see S. Peterson 1978).
30. Overgrazing, in conjunction with the drought of the 1890's, had a devastating impact on the grassland community and the towns depended on it. This impact is reflected in the following statement made by an early settler in the region.

> When we came to Arizona in 1876, the hills and plains were covered with high grass and the country was not cut up with ravines and gullies as it is now. This has been brought about through overstocking the ranges. On the Little Colorado we could cut hay for miles and miles in every direction. The Aztec Cattle Company brought tens of thousands of cattle into the country, claimed every other section, overstocked the range and fed out all the grass. Then the water, not being held back, followed the cattle trails and cut the country up. Later tens of thousands of cattle died because of drouth and lack of feed and disease. The river banks were covered with dead carcasses. (Quoted in McClintock 1921:191)

31. The "chaining" of junipers has been extended into areas where they are indigenous in order to increase the productivity of this rangeland as well.
32. One animal unit consists of a cow an its calf; a section comprises 1 square mile or 640 acres.
33. The extent of overgrazing by the Aztec Company is clear. Grazing 60,000 head of cattle on 2,000,000 acres, the Aztec Company maintained animal densities of nearly 20 animal units per section.
34. Due to their present ability to exploit underground water sources, several contemporary farmers have been able to circumvent this principal limitation and establish farming operations outside of traditionally inhabited river valleys.
35. Eagar is situated at a higher elevation than Showlow. However, due to the northeastern slope of the basin, the climatic and physical conditions encountered by early settlers in Round Valley compared more closely to those experienced by settlers in towns at lower elevations than Showlow within the western portion of the basin.
36. Alpine, the most highly-situated settlement under consideration, registered the highest proportion of non-heads of households declaring farming as their principal occupation in the 1900 census. While only 13 heads of households listed farming as their primary occupation at Alpine, 21 non-heads of households did so. This ratio contrasted sharply

with most other settlements, such as Snowflake and Taylor where the comparable figures were 34 and 6, an 27 and 2 respectively. Even Showlow registered only 2 non-heads of households compared to 14 heads of households declaring farming as their principal occupation in the same census. Nutrioso, a small farming settlement approaching the altitude of Alpine, displayed a ratio of 16 non-heads of households to 20 heads of households reporting farming as their primary occupation. Of equal significance is the young age of many of the non-household heads declaring farming as their principle occupation at Alpine. Five were between 10 and 15 years of age. The comparable figure for Nutrioso was 4. This marked disparity displayed by Alpine and Nutrioso regarding the ratio of dependents to heads of households employed in farming suggests that forcible demands were placed on farming at this altitude. Very likely, all available labor was needed to complete the full complement of agricultural operations within the restricted growing season provided. Given the variability associated with an already short growing season at this elevation, poor harvests were likely a regular occurrence for farmers at Alpine, Nutrioso, and other mountain settlements.

37. While the depth of this alluvium does not exceed 30 feet in most places (Babcock and Snyder 1947:7), its thickness in at least one location near Joseph City has been measured at 180 feet (Dames and Moore, 1973 [section 4]:155).
38. See Chapter 5.
39. See Chapter 5, footnote 28.

Chapter 5

Dam Construction

Of the several environmental conditions that imposed themselves on the early agricultural communities in the Little Colorado River Basin, none exerted a more profound impact on the productivity and stability of specific settlements than did the variation displayed in the abundance and distribution of suitable water for irrigation purposes. Significant, recurring and frequently unpredictable fluctuations in precipitation and stream flow made it imperative that each Mormon town construct viable water control systems immediately upon settlement, and that such structures be continually maintained--often at great individual and community expense.

Irrigated agriculture needed to overcome serious limiting factors before it could provide a reliable productivity from which stable communities might evolve. Because surface water flow is a function of the larger climatic regime, the availability of water for irrigation has been markedly irregular, and annual, seasonal and daily variations in water flow have combined to make the Little Colorado and most of its tributaries unreliable sources upon which to base agricultural systems. Since stream runoff figures for the basin are highly variable, storage reservoirs must be employed to impound water during periods of high runoff which can then be released during periods of low flow. By "smoothing out fluctuations in input" (H. Odum, 1971:16), such storage structures enhanced the stability and predictability of a principal environmental condition affecting agricultural productivity and community development in this variegated basin.[1]

The lower valley settlements were continually plagued by the problem of a variable water supply. As already indicated, geological limitations dictated that only diversion dams could be constructed in this subregion.[2] Agriculture along the upper Little Colorado River and near the headwaters of Silver Creek, on the other hand, was very early based on the use of storage reservoirs due to the existence of suitable geological conditions at these locations. Consequently, unlike St. Joseph, Woodruff and the other lower valley settlements, the more southerly towns were better able to adapt to the seasonal shortage of streamflow during May, June and July. They were, therefore, better poised to overcome a critical limiting factor constraining agricultural productivity and community development in the region.[3]

The continued deposit of silt in the stream bed of the Little Colorado River throughout the lower valley over many thousands of years also contributed to the difficulties encountered by lower valley settlements in their quest for viable irrigation systems. Nineteenth century Mormon irrigation system in the lower valley all suffered from the same problem: the lack of a firm foundation upon which to found diversion dams. Commenting on the difficulties experienced in dam construction along the lower river, McClintock (1921:141) aptly characterized the Little Colorado as a "treacherous stream at best, with a broad channel that wanders at will through the alluvial country that melts like sugar or salt at the touch of water." Not only did dams in this section of the basin wash away with frustrating regularity, but on more than one occasion a dam was left high and dry when the river abruptly altered its course.

The construction and maintenance of irrigation ditches also imposed a considerable burden. Labor and assessments paid on ditches were often as hard on the settlers as those spent on the dams themselves. Moreover, as much as 30% of the water diverted for agricultural purposes was lost through evaporation and seepage in these canals (cf. Bureau of Reclamation 1947:14). Consequently, when choosing a damsite, the settlers not only had to pick an optimal location for the dam itself; they also had to consider the length and route of the attending ditches.

Permanent dams were not achieved by some settlements in the region until the 1920's, and the histories of certain towns prior to that decade have largely been recalled as a succession of dam failures and reconstructions (see Peterson 1973:185-191; Tanner and Richards 1977:41-50; Richards and Westover 1964). Recurring dam failures were particularly characteristic of settlements along the lower valley of the Little Colorado River. Although dam losses occurred on the upper Little Colorado and on Silver Creek, they were considerably less frequent (see Table 5.1). The lower river settlements suffered doubly, moreover, for when the dams upstream did burst, they precipitated a chain reaction which frequently destroyed the dams along the lower river as well. Leone (1979:91) registers nearly 50 dam losses in the basin between 1876 and 1923, of which 27 occurred at St. Joseph and Woodruff alone.[4] The actual devastation sustained, however, was greater than even these figures imply. Multiple dam failures in a single year plus numerous, equally expensive near-failures--frequent occurrences not accounted for in Leone's table--increased the actual costs incurred by these early settlers.

The timing of a dam washout was critical. If a dam collapsed during the winter months, as Leone (*ibid.*) indicates that about half of them did, the loss was not necessarily ruinous to the coming year's harvest. Enough time might still have existed (provided sufficient manpower was available) to rebuild the dam prior to the beginning of the next agricultural season. If, on the other hand, a dam failed during the growing season, most notably with the onset of the summer rains, the effect on the fall harvest was likely to have been devastating. Even a dam collapse associated with the heavy runoff of the early spring might have precluded an adequate harvest prior to the first frost of that year unless an ample supply of labor was available to complete the task quickly. Such labor availability was generally beyond the capability of most settlements.

Inasmuch as irrigation systems constituted the most significant and expensive community investment in productivity among Mormon Settlements in the Little Colorado River Basin throughout the nineteenth century, local differences in the amount of resources expended on dams represents a vivid indication of the relative size of the drains on community productivity borne by specific settlements. For this reason, the present chapter will focus on the history of dam constructions among various settlements.

Table 5.1

Dam Losses among Individual Settlements
in the Little Colorado River Basin

Settlement	1876 - 1900	1876 - 1923
Lower Valley Settlements:		
St. Joseph	13	14
Woodruff	10	13
Intermediate Settlements:		
St. Johns	2	5
Snowflake	3	6
Taylor	3	6
Eagar	0	1
Mountain Settlements:		
Showlow	1	1
Alpine	0	0

SOURCE: Leone (1979:91); Little Colorado Stake (n.d.); Eastern Arizona Stake (n.d.); St. Johns Stake (n.d.); Snowflake Stake (n.d.).

The St. Joseph Dams
Tanner and Richards (1977:41) have appropriately labelled St. Joseph's efforts to build a viable irrigation system as "The Eighteen Years War",[5] during which time the settlers of this small community constructed nine dams.[6] Their account of successive dam constructions and failures at St. Joseph is perhaps the best available (see Tanner and Richards 1977:41-50) and will largely be followed here.
　　The first dam, as already discussed, was a joint undertaking, constructed in cooperation with the settlers at Obed (see Chapter 2). St. Joseph's share of the

labor on this dam and its attending ditches demanded that nearly its entire adult male population (about 50 men) work for nearly 2 1/2 months, during which time 1,208 man-days and 354 team-days labor were consumed. This dam was destroyed with the very first freshet that came down the river.

During the subsequent winter and spring, settlers at St. Joseph constructed a new dam two miles further upstream from the site of 1876 dam. They, therefore, had to add two miles to the ditch system (in addition to repairing existing ditches). Labor on the 1877 dam and ditch placed considerable stress on the settlers at St. Joseph as the entire project had to be completed by only 18 to 20 men (Porter n.d.a.:355). Despite their strenuous efforts the settlers produced only a modest harvest. When freshets finally arrived late in the season, they cut a new channel in the river around the dam, "rendering it wholly useless" (Tanner and Richards 1977:42).

The 1878 agricultural season was characterized by especially high floods, capped by an excessively wet harvest (Porter n.d.a:358), and the flood of August 22 that year remains among the highest on record. Most settlements along the Little Colorado lost their dams that year, and St. Joseph was no exception. The settlers there struggled determinedly to save the dam which they had built the previous year, but to no avail. As Nuttal (1878b) reported during his stay at St. Joseph,

> Their dam cost about $3,000 in labor, is 300 feet long, 40 feet wide, 4 feet high; also a levee at ends of dam 400 feet long, 8 feet wide and from 3 to 5 feet high. During the late floods the river changed its course, so that the dam and levees are now worthless; the water ditch from the dam is some 9 miles long and much damaged by the late rains.

Eight hundred additional man-days of labor, not counting the use of team labor, were consumed by the settlers at St. Joseph while attempting to save their dam (see Little Colorado Stake:9/31/1878). Yet, by the close of the 1878 season, "both dam and ditch were in worse condition than when work commenced on them the preceding February" (Porter n.d.b.:5).

In 1879, work proceeded on a new dam which was located an additional mile up the river. The ditch extending from this dam was very expensive. Not only was it a mile longer, but it also required several deep cuts through protrusions in the landscape. This dam lasted for two years. The cost of dam and ditch maintenance for 1880 was recorded at only $300 (Porter n.d.a.:360), was considerably less than the amount expended during any previous year.

The next dam, constructed in 1881, was located back downstream at the site of the original 1876 dam. While no clear reasons are given for investing in a new dam at the original location, the costs of the longer ditch (both in terms of the labor expended on maintenance as well as the water lost through shrinkage and evaporation) may have proved excessive. In any case, the 1881 "high-dry" dam was a radical departure from previous structures.

Dam Construction

The idea was to build a large enough earth dam across the river channel to force the stream over its banks and form a new channel on a rocky area. The earthen dam would have to be high enough to turn aside the largest flood, and would need to be protected on the exposed end where the water would pass so it would not erode away. To protect the face of the exposed earthen dam a crib was built of logs bolted together and filled with rock. If the settlers were successful in turning the stream on to the solid rock, it would not be too difficult to control it, since there would be a solid base. All the settlers believed that their failures in the past were due to the sandy bottom of the river. (Tanner and Richards 1977:43-44)

By March 16th, work had progressed far enough to begin closing the last gap in the dam which had been left open to permit water to pass through during construction. By three that afternoon this last gap was closed and the dam completed. On March 20th, just four days later, a freshet of muddy water originating from northern tributaries swept down the Little Colorado and carried the dam away (Tanner and Richards 1977:44). The estimated cost of this destroyed dam was $6,000 in man and team labor (Porter n.d.c.:83).[7] A second dam was apparently constructed that same year, because on November 26, 1881 St. Joseph was reported to have lost another dam (*ibid.*).

After the devastating loss of such an expensive dam, the settlers at St. Joseph tried to minimize their investment the following year. The 1882 dam was constructed further downstream than any of its predecessors and was viewed as a temporary structure which would wash away with the duration of the summer rains. While the dam's downriver location allowed for a shorter irrigation ditch, a completely new canal had to be constructed from this previously unused site which ultimately consumed more labor than the dam itself. Labor on the new dam and ditch was supplemented by workers from as far away as Snowflake and St. Johns. Although many came as volunteers,[8] a total of $1,000 was paid in compensation (Tanner and Richards 1977:45).

While the 1882 dam was of simple construction compared to its predecessors, the ditch was another matter. Because the downstream location of the new dam required that the attending ditch cross several washes before reaching the fields, troublesome wooden flumes had to be incorporated into the ditch's construction. The 1882 ditch ruptured on several occasions and demanded considerable repair. One early pioneer noted that "the 4th of July passed without any celebration because of so much work" (quoted in Porter n.d.b.:6). Despite their indefatigable efforts, the settlers at St. Joseph could only boast a "below average" harvest (Tanner and Richards 1977:45). At this time the local bishop estimated that the town had expended a cumulative total of between $30,000 and $35,000 in labor on dams without arriving any closer to a solution (Porter n.d.c.:84; see also Tanner and Richards 1977:45).

No record exists of any expensive irrigation construction during 1883, perhaps because this was not a particularly wet year (see Porter n.d.c.:86).

However, during March of 1884, heavy floods inflicted extensive damage, including the complete destruction of the 1882 dam. Construction of a new dam was begun almost immediately. Fearing a total crop failure due to the late beginning of dam construction, some planting was also done in the mountains above Heber, a settlement founded by colonists from St. Joseph the previous year (Porter n.d.b.:6).

The 1884 dam represented another novel approach to dam construction on the part of settlers at St. Joseph. With the help of the Atlantic and Pacific Railroad, 22 pilings between 16 and 20 feet in length were driven in two rows across the riverbed at the location of the original 1876 dam, and a structure composed of timber, brush and rock was erected against these piles (see Porter n.d.b.:6). Dam construction began on April 8th and water was turned into the ditches on May 18th, making this one of the most rapidly completed levees on the Little Colorado (see Tanner and Richards 1977:46). While the pilings themselves withstood the ravages of the river, the materials adjoining this foundation were chronically torn loose by heavy flooding, and major repairs were required each spring. Rejuvenated by annual repairs, the 1884 dam served the settlement until it was entirely swept away by the heavy flooding which occurred during the fall of 1887. Survival for four complete agricultural seasons made the 1884 dam the most successful St. Joseph dam to date.

Little information exists on the dams used during 1888, 1889, and 1890, indicating that they were not very substantial structures. Porter (n.d.c.:89) claims that while they were superior to sand dams, they were constructed with an expectation of surviving only one season. Located at the site of the 1882-1883 structure, these dams shared the problem of troublesome flumes and ditches which plagued the settlers previously (Tanner and Richards 1977:46). At the same time, "major sections of the dam, if not the entire structure, were taken away with disheartening regularity" (ibid.). Consequently, while not requiring the major initial expenditure associated with certain previous dams, the dams of 1888 through 1890 are not likely to have represented insignificant investments either.[10]

The next major dam construction was undertaken in 1891. It took three years to complete, and represented the most significant incorporation of total community effort since the days of the early United Order colonization. This dam marked a return to the "high-dry" type dam, and was built near the site of the original 1876 dam. At the existing wage rates, the final cost of this dam has been variously given as $5,000 and $6,000 (see Porter n.d.c.:91; McClintock 1921:142; Tanner and Richards 1977:47).[11] However, one cost not included in these figures was the relative neglect of farming that necessarily resulted during the three years in which this dam was under construction (see Porter n.d.b.:7).

So substantial was the undertaking initiated by this small settlement that an unofficial return to the United Order working and eating arrangements was established (see Tanner and Richards 1977:47). A "dam house" was erected for the men working at the construction site, and the women of the settlement took turns, two-by-two, cooking for the men working on the dam. Between 10 an 25

Dam Construction 127

"men and boys" labored on the dam, supplemented by workers from Snowflake and St. Johns (*ibid.*). Porter (n.d.b.:7) describes the dam as follows:

> It was across...(the)...deep quicksand bed, where no substantial footing could be reached, that the high part of the dam of 1891 was built. A "crib", made of heavy timbers securely bolted together and to the underlying stone and sufficiently high to be well above the flood waters when they were turned over the stone table to the northward, was placed as near the edge of the quicksand-filled canyon as was deemed practicable. This crib was loaded with stone and gravel to give it weight and stability; and to the southward from it, the "high-dry" dam was constructed to the more elevated walls beyond the river. To the northward of the crib was the spillway, 191 feet in width, through which the flood waters were forced. Across part of this spillway, insufficiently high to turn a low river stream into the mouth of the canal, some heavy timber construction was placed, also securely pegged to the stone beneath, and reinforced by a sealing mass of stone, gravel, and earth along its upper edge.

By June of 1891, water was running in the canals. However, for the next three years the settlers at St. Joseph were forced to alternate between working on their farms and laboring on the dam, hoping all the while that no unusual flood would destroy the structure before it was completed. The dam was finally dedicated amid considerable celebration on March 10, 1894 (see Porter n.d.b.:7).[12]

The 1894 dam was not the final dam at St. Joseph. Although it withstood for 29 years, considerably longer than any of its predecessors, the entire structure was swept away in 1923 when heavy runoff caused the collapse of Lyman Dam above St. Johns and unleashed a torrent of water unlike anything ever released in this stream. A new and final dam was constructed at the same site in 1924. This dam was 200 feet long, 12 feet thick at the base, 10 feet thick at the top, 26 feet high and contained 12,500 cubic yards of earth plus considerable rock work (see Porter n.d.c.:93-95; Tanner and Richards 1977:48). The most significant feature of this new dam was the inclusion of cement, which because of its great expense had not been previously used in dam construction in the basin. Three hundred and twenty-five cubic yards of cement was used in the building of this final structure (*ibid.*), which was made possible through a $3,000 appropriation by church headquarters in Salt Lake City (Tanner and Richards 1977:50). The cost of the cement raised the total cost of the 1924 dam to about $16,000 (Porter n.d.c.:95).

As the expense, instability and generally poor returns associated with agriculture at St. Joseph have demonstrated, irrigated farming did not represent a viable adaptive strategy upon which to build a firm community base at this location during the nineteenth century. Continually plagued by limited resources, settlers at St. Joseph could barely surmount the drains imposed upon them by a variable and unpredictable environment. Consequently, they became increasingly dependent on strategic subsidies derived from external sources. The difficulties inherent in attempts to establish viable agricultural settlements in the lower valley of the Little Colorado River are clearly illustrated by the fact that St. Joseph was the only original colony to survive. The point is made doubly clear by an examination of the history of dam construction at Woodruff, another well-documented sequence. Between 1878 and 1919, Woodruff built 13 dams and suffered the destruction of 12 (see Fish n.d.a.:34-37; Peterson 1973:185-191).

The Woodruff Dams

Unable to complete a dam across the Little Colorado in 1877 due to their limited numbers (see Chapter 2), the settlers at Woodruff expended considerable labor in dam construction the following year. Investing 390 man-days and 227 team-days labor,[13] the handful of settlers at Woodruff erected a dam which was 125 feet long, 50 feet wide and 22 feet high,[14] with expectations running high of turning water into the fields by early May. Before the structure was completed, however, a heavy flood swept the dam away (Little Colorado Stake:5/24/1878). With the loss of this dam and the subsequent failure of the 1878 harvest, only 3 families remained at Woodruff (Peterson 1973:187), and the spring of 1879 found the settlement nearly deserted.

During the summer and fall of 1879 several additional families settled at Woodruff, and in November of that year construction began on a new dam which was completed the following May. Owing to the generally low flow of water in the Little Colorado during the late spring and early summer, the increasing water level behind the Woodruff dam in 1880 threatened the crops downstream at St. Joseph. Following an urgent appeal by farmers at St. Joseph, a hole was cut in the Woodruff dam. "This was a damage of about $500.00 and blasted their hopes of a crop for that season" (Fish n.d.a.:35).

A flood during September of 1881 was reported to have destroyed a considerable portion of the 1880 dam (*ibid*.), and in January of the following year work was begun on the construction of a new dam (*ibid*.). Although the labor on this new dam was heavy, owing to the small size of Woodruff's population, 1882 displayed positive results, as Fish (*ibid*.) reports:

> The work was continued during the winter and spring with about an average of 7 men and in May the water was again brought out. A small field was fenced and about 70 acres of land put in mostly to corn and vegetables. This yielded well and the people were greatly encouraged.

During the winter of 1882-1883 the dam was reinforced and nearly 400 acres were planted, with the prospects appearing favorable that spring for a successful harvest (Fish n.d.a.:36). On July 26, however, another devastating flood swept down the Little Colorado destroying most of the dam in its wake. Fish (*ibid*.) reports that a leak started around the west end of the dam. This increased rapidly and not withstanding strong efforts being made to stop it, the breach soon widened and the bank and the end of the dam were soon cut away to a depth of 25 feet and 100 feet wide. This was a very heavy blow to the place and many turned away from the sight feeling to give up the enterprise.[15]

Construction began on the next dam on January 2, 1884 (*ibid.*) and continued to near completion when a freshet completely washed it away. A call was issued to members of the various wards at the next stake conference to donate one week's labor towards rebuilding of the Woodruff dam. Work on the second dam of 1884 commenced on June 2nd, very shortly after the destruction of the preceding structure. An average of 45 men labored on the dam at any one time, and $3,000 in labor is estimated to have been donated (*ibid.*). Just as this second dam was also nearing completion,

a heavy body of water came down cutting over it and it soon went out leaving it about as it was one year before. This was a heavy blow to the people for the season had been quite dry and the little grain that had been put in had completely dried up and was not worth harvesting. (*ibid.*)

Settlers at Woodruff were able to survive the winter of 1884-1885 because of their participation in the construction of the new building to house the ACMI store there (see Chapter 6). Local church leaders awarded the contract for the construction of this two-story building to the residents of Woodruff so that they might be supported until a new dam could be completed. By the close of 1885, the Woodruff dam had been rebuilt and a permanent structure finally appeared to have been achieved. No dam failures were reported for several years.

On February 21, 1890 this growing optimism was quickly displaced as an extremely large flood on Silver Creek destroyed the Woodruff dam in a matter of minutes.[16] Church authorities in Salt Lake City appropriated $3,000 from local tithing funds to subsidize the reconstruction of the Woodruff dam (Fish n.d.a.:37).[17] At the same time, $1,500 in relief was appropriated to the residents of Woodruff by the Arizona Territorial Legislature (Peterson 1973:190). A small population (and thus a limited labor supply) caused work on the dam to proceed slowly. Although they were able to turn water into their fields late in the season, the settlers at Woodruff were forced to leave their dam in an unfinished condition. Consequently, when a second flood passed on November 8, 1890, all of their labor went with it, and several persons made plans to abandon the settlement (Fish n.d.a.:37).

Church leaders again assisted in directing the reconstruction of the Woodruff dam, and donated labor from the surrounding towns was forthcoming (see Peterson 1973:188) as little of the previously extended provisions remained. This dam, flagged by stone slabs (Peterson 1973:185) and costing perhaps $11,000 (see Woodruff Irrigation and Recreation Project:i), held for 14 years until Zion's Dam above Hunt burst on August 26, 1904 under the pressure of heavy rains (Fish n.d.a.:37). So immense was the torrent of water released by the collapse of Zion's Dam that the residents at Woodruff were forced to abandon their homes and seek shelter at higher ground (*ibid.*).

During November of 1904, yet another appeal was made by the stake for assistance in reconstructing the disabled Woodruff dam. Church headquarters contributed $500 in cash and the Snowflake, Taylor, St. Johns and Showlow wards were asked to contribute an additional $1,500 jointly (*ibid.*). The settlers at Woodruff were forced to farm without irrigation water for two full seasons as they were unable to turn water onto their fields until August 2, 1906 (Fish n.d.a.:37). This twelfth and last dam on the Little Colorado was constructed entirely of stone, and lasted until 1915 when Lyman Dam (which impounded the largest reservoir in the entire basin) faulted under the pressure of heavy stream flow and released a flood of water which swept this dam away (see Peterson 1973:185-186).[18]

After repeated failures along the Little Colorado, the construction effort turned to Silver Creek. Winding its way through moderately deep canyons, Silver Creek possessed the rocky foundation upon which a permanent dam structure could more securely be built. Construction of a dam on Silver Creek also offered an escape from the "silts and minerals of the Little Colorado as well as its fierce floods" (Peterson 1973:186). A permanent diversion dam was finally completed in 1919, but only after the State of Arizona appropriated $10,000 and the Mormon Church contributed $22,500 towards its final cost of $85,000 (see Peterson 1973:188). This was an expensive irrigation system. Beyond the meager resources of this small settlement, it incorporated technical problems of ditching along canyon walls as well as the placing of flumes across canyon crevasses.

> At places, the canal was hung from the cliff above, at places it was chiseled through solid rock, and at other places it took the shape of flumes across the river or wash. (Woodruff Irrigation and Recreation Project n.d.:i)

Although Woodruff's struggle with water is locally viewed as ending subsequent to the completion of the 1919 dam, in fact, the struggle continued. The flumes leaked regularly, demanding constant repair, and water losses in the ditch became excessive (see Peterson 1973:186; Woodruff Irrigation and Recreation Project, n.d.:i). These heavy costs eventually led to the abandoning of irrigation by means of the Silver Creek dam. Farmers turned instead to the pumping of water from underground sources.[19]

The history of its struggle to control the Little Colorado River, like that of St. Joseph, reveals "the fineness of the balance in which Woodruff's existence hung. Closely circumscribed by natural conditions, it barely escaped joining the

Little Colorado roster of ghost towns" (Peterson 1973:187). Varying up to a peak of 30 families, Woodruff's population waxed and waned depending on the status of its dam. Each successive dam failure triggered an exodus of settlers searching for more optimal locations. Continually plagued with a labor supply inadequate to complete the construction of another dam, a core of a very few families persisted in recruiting settlers to augment their sparse numbers. However, each successive dam failure testified to the cost and unreliability of farming in this valley. In the final analysis, the settlers at Woodruff survived only because of externally derived subsidies which offset the failures of their agricultural system. Besides the numerous and substantial contributions of money, labor and produce received from nonlocal sources for the explicit purpose of subsidizing dam reconstruction, the continued occupation of this valley was largely made possible by individuals continually seeking employment elsewhere in the basin. An inventory of the population at Woodruff after the failure of the 1904 dam revealed that although 33 families inhabited this small valley, 10 were headed by widows or aged men while the men of 20 families were scattered in employment throughout the basin (see Peterson 1973:190). Settlers at Woodruff were unable to channel sufficient resources into agricultural production due to the drains imposed on them by an unstable river. They, therefore, operated a costly and largely ineffectual agricultural system. Consequently, like St. Joseph and the other lower valley settlements, Woodruff never achieved sufficient prosperity (i.e., a substantive and reliable surplus) from which to generate community development.

St. Johns Dams

Although dam failures did not occur at St. Johns with the regularity displayed at either St. Joseph or Woodruff, such infrequency was partially compensated for by the scale of the irrigation systems there and by the size of the losses caused by their destruction. The dams constructed by the settlers at St. Johns dwarfed all others in the basin. Consequently, when they collapsed they unleashed massive torrents of water that destroyed everything in their path--including, generally, the dams at Woodruff and St. Joseph which might otherwise have withstood the stresses of the swollen river.

The initial irrigation works at St. Johns antedated Mormon settlement at this location and, compared to subsequent structures, comprised very rudimentary water control systems (see Fish n.d.a.:49). Built in the early 1870's by the indigenous Hispanic population, the original irrigation system at St. Johns consisted principally of a small dam and three irrigation ditches designed only to conduct water upon land immediately adjacent to the Little Colorado River channel.[20] With the dramatic increase in population at St. Johns following the immigration of hundreds of Mormon settlers between 1880 and 1885,[21] irrigation by diversion alone could not accommodate the growing agricultural demands for water. Due to the generally diminished supply of water in the Little Colorado during the late spring and early summer, open hostility developed between Mormon settlers and the indigenous Hispanic population over conflicting claims to Barth's rights to water from this stream.[22] In an attempt to overcome this seasonal shortage of water and at the same time accommodate the rapid influx of settlers, the Mormon population at St. Johns constructed two storage reservoirs during the early 1880's. One reservoir adjoined St. Johns on the north, covering an area of about 60 acres, while the second lay 1.25 miles south of the town and extended over 125 acres (Fish n.d.a.:51). The combined cost of these two dams with their attending canals exceeded $4,000 (see *ibid.*).

In order to enhance agricultural productivity, the St. Johns Irrigation Company was formed which undertook the construction of a large dam six miles south of the town.[23] Slough Dam, as this structure was named, was reported to be nearing a "Christmas completion" in August of 1894 (Jensen n.d.d.:8/9/1894), and during the following three years this finished dam was reported to have impounded a substantial capacity of water for irrigation purposes. Slough Dam apparently underwent substantial repair and expansion during most of 1900 and likely into 1901, as several reports included within the Minutes of the St. Johns Stake Conferences during this time period referred to a large dam under construction also located six miles south of the town. One report dated February 10, 1900 indicated that much work was in progress on a large reservoir six miles upstream which when finished would be 40 feet high, 150 feet long and 150 feet thick at the bottom, and which would flood 1800 acres of land.[24] This structure was built of rock on the lower side tied together with cedar, with the upper or inside facing the water made of dirt. It was expected to permit the cultivation of an additional 4,000 acres of land (*ibid.*). No such detailed descriptions are available for the original Slough Dam. Furthermore, Leone (1979:91) indicates a possible dam loss at St. Johns in 1899, which would account for such a large reconstruction effort being expended in 1900.

Work on this second Slough Dam required strong community involvement. Large numbers of people participated in its construction, and labor on the dam assumed a "united effort" (Jensen n.d.d.:12/1900). By December of 1900, the volume of water impounded behind the second Slough Dam was greater than that stored during any previous year at St. Johns, forming a lake nearly 3 miles long. Even at this volume, the water had an additional 10 to 15 feet to climb before it reached the top of the dam at its existing stage of construction (*ibid.*). However, this enlarged Slough Dam collapsed in 1903, producing several years of population decline and economic depression at St. Johns (see McClintock 1921:182).[25]

A renewed construction effort was begun in 1910 when the residents at St. Johns (in association with a Denver, Colorado development company)[26] jointly incorporated into the Lyman Irrigation Company for the purpose of building a large reservoir 12 miles south of the town (see Berry 1910). The Colorado company and the settlers at St. Johns each provided one-half of the necessary capital to underwrite this substantial project,[27] and the Mormon Church appropriated $5,000 to aid the residents at St. Johns in raising their share of the investment

(McClintock 1921:183). Completed only after an expenditure in excess of $200,000, the construction of this dam proved defective. In April of 1915 the structure collapsed, killing 8 persons and causing extensive damage to flooded farm lands throughout the region (*ibid.*).[28]

The Lyman Company was subsequently reorganized, and by 1918 another $200,00 was spent. However, St. Johns still lacked an adequate water storage facility. An appeal was then made to the State of Arizona. The State Loan Board advanced about $800,000 to the residents of St. Johns, holding mortgages on their land and on the dam as its security (*ibid.*).[29] This second Lyman Dam, which still stands, was completed in 1920.

While not experiencing the near-annual setbacks sustained by St. Joseph and Woodruff, St. Johns felt the devastation of successive dam failures. Moreover, the sheer size of the dams at St. Johns, the volume of water that they harnessed, and the substantial community investment that each represented made subsequent losses increasingly difficult for this community to endure. As in the cases of both St. Joseph and Woodruff, a final solution to St. Johns' water control problem was achieved only through the application of strategic subsidies received from external sources.

Irrigation at Snowflake and Taylor

When the early Mormon pioneers purchased Stinson's claim along Silver Creek, he had about 300 acres of land under cultivation (McClintock 1921:164). Mormon farmers were to increase the amount of irrigated land in this vicinity to about 2,200 acres.[30] While not without its setbacks and hardships, irrigated agriculture in the Snowflake-Taylor area experienced a steady amelioration through the expansion and consolidation of existing irrigation systems. Snowflake and Taylor did not suffer the continual stresses which drained St. Joseph and Woodruff; nor did they sustain the occasional devastations which haunted St. Johns. For the most part, each successive investment in the water control system serving these two interdependent settlements enhanced their future capacity for irrigation rather than merely allowing for the recovery of lost ground.

Irrigation at both Snowflake and Taylor was first accomplished through the direct diversion of water from Silver Creek. During 1878, five men built a dam with a ditch on the west side of Silver Creek at Taylor. A second canal was constructed on the east side of the river the next spring to accommodate the influx of settlers from Woodruff following the failure of their dam (see Fish n.d.a.:43). The settlers at Snowflake also relied solely on water diversion initially. In the spring of 1879 heavy work was reported being performed repairing ditches which had been severely damaged (in some places completely destroyed) by the heavy rains and high waters of the previous fall (see Fish n.d.a.:40; LeVine 1977:23). With increasing immigration, diversion alone quickly proved inadequate to meet the growing demand for water, and dissention developed between the settlers of Snowflake and Taylor regarding competing claims to irrigation water.[31] Except for this early conflict, the development of irrigation along Silver Creek proceeded with the complete integration of the water control systems serving these two towns.

In an effort to enhance the farmable acreage within their adjoining valleys, farmers at Snowflake and Taylor constructed several small storage reservoirs prior to the early 1890's. One of these reservoirs, constructed in 1887, was described by Fish (n.d.a.:22).

> In the latter part of January a reservoir was surveyed (by Joseph Fish) between Snowflake and Taylor on the west side of the creek. This was made and completed (so far as intended for the present) before the first of March by the people of Snowflake. There is about 4550 (cubic) yards of earth work in the bank besides rock and brush which is put in for protection. It is estimated that this will hold enough water to irrigate 1,000 acres over once. (parentheses in the original)

Snowflake and Taylor each lost their dams during the heavy floods of 1890. However, construction began on the Flake Ranch Reservoir immediately following the formation of the Snowflake and Taylor Irrigation Company in 1894 (see Bureau of Reclamation 1947:2, 34). Two additional reservoirs were added to the system in 1898 (*ibid.*). Another dam, replacing the two extant Snowflake dams, was built between Snowflake and Taylor in 1899 (LeVine, 1977:113).

Finally, work began in 1906 on the Daggs Dam and Reservoir located 10 miles south of Taylor. Daggs Dam, which impounds 2,500 acre-feet of water, was built to replace several previous dams and provide the bulk of the irrigation requirements for farmers in the Snowflake-Taylor area. When finished, this dam stood 65 feet high, 230 feet thick at the bottom, 14 feet thick at the top, 60 feet long at the bottom and 330 feet long at the top. By 1909 Daggs Dam had reached a height of 33 feet, and construction was finally completed on May 23, 1914 at a total cost of $40,000 (*Snowflake Herald*:6/6/1914; see also Levine 1977:114),[32] with all of the labor performed by farmers working out their assessments (Bureau of Reclamation 1947:34). While future repairs, reinforcements and enlargements of Daggs Dam occurred,[33] the dam was never lost, and Daggs Reservoir served as the principal source of irrigation water for these two towns until it was converted to recreational use in the 1960's. At this time, farmers at Snowflake and Taylor turned to using water pumped from the Coconino sandstone (see Abruzzi 1985).

While irrigated farming along upper Silver Creek was not without its difficulties,[34] an examination of the history of dam construction along this tributary presents a very different picture from that along the Little Colorado. The most arduous and persistent demands imposed by irrigation in the Snowflake-Taylor area--and these were mentioned often--were those associated with the maintenance of irrigation ditches. One reference to problems with ditches has already been made. Fish (n.d.a.:67) indicates that ten years later ditch repair still presented a considerable drain.

Dam Construction

The labor on ditches in this stake has been enormous, at Snowflake this year the tax was a trifle over $3.00 per acre. Paying out this amount every year per acre on dams and reservoirs tells upon the people and they begin to think it is too much of a burden to carry.

The maintenance of dirt ditches was a constant problem, particularly during years of heavy rainfall. Added to this cost was the already mentioned loss through evaporation and seepage of nearly 30% of the water diverted for irrigation. Maintenance and seepage problems in Snowflake were never effectively solved until the 1960's when at least the primary canals were lined with cement (see LeVine 1977:118).

However, Snowflake and Taylor were not alone in their struggle to maintain irrigation canals or in the losses of water they suffered during transit. While the assessments for ditch repair in these two towns were high, they were comparable to those levied elsewhere in the basin (see Fish n.d.a.:67). Canal maintenance was equally problematic for other towns as well, especially those in the lower valley where the soil through which irrigation ditches traversed has been characterized as "sugary" (Peterson 1973:180; see also McClintock 1921:141). Furthermore, while struggling with their ditches the farmers at Snowflake and Taylor did not have to contend with the even greater drains imposed by repeated dam failures. Significantly fewer dam losses occurred at Snowflake and Taylor than at either Woodruff or St. Joseph, and the scale of these losses was insignificant compared to that endured by St. Johns.

CONCLUSION

The preceding pages clearly illustrate the extent of the drain imposed on those settlements which sustained recurring dam failures. The costs incurred encompassed not only the reconstruction of damaged or demolished dams, but also the destruction wrought upon ditches, crops and, ultimately, on the fields themselves. Successive, poor harvests in association with repeated dam losses imposed an extraordinary hardship on individual settlers, undermining their ability to continue the struggle with the Little Colorado. Population size in those settlements most subject to dam failures either fluctuated markedly or declined absolutely. Such instability derived from the persistent inability of these settlements to recruit and retain sufficient manpower to effectively counter the river's variability and power. Consequently, those settlements most affected by the difficulties attending irrigation in the basin were able to endure only because strategic inputs of labor and other resources were forthcoming from non-local sources.

Externally derived inputs of labor, produce, materials and money comprised indispensable subsidies without which certain settlements likely would not have survived. Without the judicious appropriation of such resources, St. Joseph and Woodruff (containing among the smallest populations in the basin and enduring the most adverse conditions accompanying irrigation) would surely have followed their neighboring settlements to extinction. While the Mormon settlement at St. Johns is not likely to have dissolved in the absence of specific subsidies, the population of this town did wax and wane with the status of its dam, and the productivity of its agriculture was central to its regional pre-eminence.

Despite the fact that the problems attending irrigation were of central concern to all nineteenth century Mormon settlements in the basin, specific data relating to water control for individual towns has been unevenly preserved. While the quantity and quality of available information bearing on the costs attending successive dam failures and reconstructions at St. Joseph and Woodruff are quite impressive considering the frontier context in which they were recorded, this same level of reporting was not maintained at other settlements. Although at least a minimal amount of data is available regarding the irrigation systems at St. Johns, Snowflake and Taylor, almost no information has survived which illuminates the struggle with water control systems waged by the more remote settlements (including Eagar,[35] Showlow and Alpine) situated at higher elevations. This variation in data availability reflects, in part, the lesser direct involvement in the affairs of these more remote settlements of church leaders overseeing the colonizing effort (cf. Peterson 1973:36), as well as the different concern among individual settlers for recording the procession of their historical venture. However, such differential data availability also reflects the relative intensities of the struggles waged. Those settlements which preserved the most complete record of their labor on water control systems, attending to every detail, are precisely those towns for which the problems surrounding dam failures assumed the principal theme of their local histories and generated wider church interest. At the same time, those settlements which sustained the fewest dam failures have not been overly concerned with recounting the detail of these infrequent events.

One final comment regarding the frequency of dam failures needs to be made. Besides demonstrating a distinct difference in the incidence of dam failures among individual settlements, Table 5.1 testifies to the significant distinction that existed between settlements in the lower valley and those located elsewhere in the basin. The specific conditions outlined in Chapter 4 placed a substantially higher stress on irrigation systems in the lower valley, undermining the viability of agricultural settlements in this area. Indeed, of the six settlements founded along the lower Little Colorado between 1876 and 1878, only two survived the decade following their colonization. No other sub-region matched the extinction rate achieved in the lower valley; every town established in river valleys at intermediate elevations survived, and those highland settlements which were abandoned were deserted for reasons unrelated to irrigation.[36] Furthermore, while settlements in the lower valley displayed marked instability in both population size and agricultural productivity, towns at intermediate elevations, most notably Snowflake and Taylor, experienced the least fluctuations in these critical community variables.

As this and preceding chapters have stressed, the Little Colorado River Basin presented early Mormon pioneers with distinct spatial and temporal variation in resource availability. While temporal resource variability--most notably that associated with the availability of water and length of the growing season--imposed distinct productive restrictions within specific habitats, the region's spatial diversity offered Mormon settlers an opportunity to circumvent local habitat limitations. Although early Mormon pioneers quickly recognized the adaptive advantage of a multihabitat exploitative strategy, their attempts to integrate resource flows from distinct habitats into a unified resource-flow system were not equally successful. The initial United Order settlements' attempts at multihabitat exploitation failed. A successful multihabitat resource-flow system was only achieved at a later date with the development of a regionally coherent system of tithing redistribution. The failure of the conjoint enterprises underscores the difficulties that temporal resource variability imposed on the lower valley settlements. The differential success of these two systems complies with Ashby's (1956:206-212) Law of Requisite Variety and with the principles of energy flow in ecological systems. Because the development of an effective system of resource redistribution was central to successful Mormon settlement in the region, the following chapter will examine local Mormon efforts at multi-habitat exploitation and provide a substantive explanation for their differential success.

NOTES

1. Irrigation reservoirs constitute the passive storage of potential energy in agricultural systems (see H. Odum 1971:38).
2. St. Joseph did construct one supplementary storage reservoir in 1902 (see Tanner and Richards 1977:94). However, due to its small size and rapid silt accumulation, this reservoir never served as a significant component of this town's agricultural system.
3. Besides providing a reliable supply of water, reservoir storage serves to reduce the velocity of water flowing through irrigation ditches. By permitting the settling of suspended sediment in water during storage, reservoirs also serve to reduce silt deposition on irrigated soils. Whereas silt in the St. Johns area became deposited largely behind the Lyman Dam (gradually reducing its storage capacity), this sediment regularly found its way onto the fields at St. Joseph.
4. Leone's figures do not include dam losses at Sunset, Brigham City, Old Taylor and Obed prior to the dissolution of these settlements.
5. Actually this war lasted 48 years. Building their first dam in 1876, the settlers at St. Joseph did not obtain a permanent dam until 1924. However, the dam completed in 1894 lasted for 29 years and may be considered as a more or less permanent structure.
6. Claiming only 9 dam failures for St. Joseph significantly understates the demands imposed on this small settlement, as this figure only accounts for the complete washouts of irrigation structures. On numerous occasions destruction to the dam, while not constituting total devastation, was so severe that little difference can be discerned between the cost of repairs following such partial damage and those attending a complete collapse.

138   Dam That River!

7. Following the ratio of human to team labor expended on the 1876 dam (see Chapter 2), the 1881 dam cost this small settlement of 110 persons 2,320 man-days labor and 906 team-days labor.
8. The use of voluntary labor from other towns to aid in dam reconstruction was a common feature among Mormon settlements in the region. The donation of labor in dam repair was an acceptable means by which inividuals in this cash-poor region could meet their tithing obligation to the Church (see Chapter 6). Access to voluntary manpower from neighboring Mormon towns was indispensable to the survival of settlements in the lower valley. Not only did they experience the most incessant dam failures, they also contained among the least adequate supply of labor.
9. The settlers at Woodruff were also plagued with ungovernable flumes in their attempt to obtain irrigation water from Silver Creek. The difficulty that this method of water conveyance posed underlay the perseverence that Woodruff residents displayed in their continued construction of dams along the Little Colorado (see below).
10. One pattern associated with the history of dam construction at St. Joseph is the vacillation between investments in major structures consuming considerable inputs of labor and the reversion to more disposable dams requiring increased annual repairs but only limited initial investments. While continual repairs were expensive, the lack of a large initial outlay represented less of a commitment of a particular dam site and structure, thus maximizing the settlement's flexibility in its response to an unpredictable river. The switch to less costly dams followed the destruction of the 1881, 1884 an 1894 dam structures.
11. Porter states that the current wage rates were 22 2/9 cents per hour for a man an 16 2/3 cents per hour for a team. While no clear indication is given, 10 hours likely represents a full day's labor, making the man and team-day values $2.22 an $1.67 respectively. Assuming a ratio of human to team labor equivalent to that employed in the 1876 dam, at a total cost of $5,000, the dam completed in 1894 would have consumed 1,742 man-days an 678 team-days labor. At a cost of $6,000 the comparable figures would have been 2,090 man-days and 814 team-days labor.
12. Noting at the time that the population of this settlement consisted of only 15 families, Andrew Jensen, the Church Historian, named St. Joseph "the leading community in pain, determination and unflinching courage in dealing with the elements around them" (quote in McClintock 1921:142).
13. At the wage rates of $2.00 per day for human labor and $1.50 per day for team labor the 1878 dam cost $1,120.50. Consisting of rock, brush and dirt most early dams at Woodruff entailed labor investments of about $1,000 (see Woodruff Irrigation and Recreation Project n.d.:i).
14. The labor expended constructing the 1878 dam was invested by a population of only 15 men, 15 women, 25 boys and 20 girls (see Nuttall 1878b), comprising a total of 75 persons in 13 families. Of the 45 children listed, 23 were under 8 years of age.
15. Most settlers at Woodruff were able to remain at this location, having obtained the supplies needed to sustain them until the next harvest from the recently-opened Arizona Cooperative Mercantile Institution store. See Chapter 6 for a discussion of the ACMI.
16. The location of all but the last of Woodruff's dams (below the confluence of Silver Creek and the Little Colorado River) made them vulnerable to exceptional flooding on either stream. Being a particularly wet year, 1890 witnessed the collapse of dams at both Snowflake and Taylor, which in turn, precipitated the destruction of dams at Woodruff and St. Joseph.

17. Most of this tithing appropriation consisted of produce from the Snowflake and St. Johns stakes to provide food for the settlers at Woodruff while they labored on their dam.
18. So great was the flooding caused by the collapse of Lyman Dam that the town of Woodruff was also flooded (Woodruff Irrigation and Recreation Project n.d.:i) See footnote 28 of this chapter for a discussion of the damages resulting from the collapse of Lyman Dam in 1915.
19. Due to the increased pumping of sub-surface water in the basin during recent years, the residents of Woodruff have encountered problems with the water available in their wells. Both the level and the quality of this water have decreased significantly, generating a movement among them to solicit federal, state and county assistance to recondition the ditches leading from the Silver Creek dam (see Woodruff Irrigation and Recreation Project n.d.). However, this project must contend with a reduction in the flow of water in Silver Creek caused by the same increased exploitation of sub-surface water sources (see Abruzzi 1985).
20. Demand for irrigation water prior to the Mormon influx into St Johns was not great, as exploitation of the landscape was generally extensive in nature. Sheepherding was the principal productive activity of a majority of the Hispanic population, and Solomon Barth, who controlled most of the water rights at this location on the river, farmed only a few hundred acres of land (see McClintock 1921:178).
21. From the time of their settlement during the winter of 1879-1880 until the spring of 1885, the Mormon population at St. Johns increased to 672 persons, largely as a result of active church recruitment. Following the settlement of St. Johns, most of the subsequent Mormon immigration into the Little Colorado River Basin was directed to this location. Fish (n.d.a.:53) reports that by 1884 about 250 men had been called to settle at St. Johns, although only about 140 remained there. The difficulties experienced by Mormon settlers at St. Johns, including considerable political harassment (see Chapter 7), inhibited many from settling permanently at this location. From its peak in 1885, the Mormon population at St. Johns declined steadily throughout the remainder of the nineteenth century (see Table 2.6).
22. In 1886, arbitrators representing each group negotiated a compromise in which the Mormons were allowed 60% of the river's capacity, with the Hispanic population obtaining rights to the remaining 40%. This conflict was soon alleviated when the construction of large storage reservoirs generally eliminated the seasonal water shortage during May, June and July.
23. The exact year in which construction began on this dam is not clear. Fish does not mention the dam's construction at all, and its existence is not recorded elsewhere until the dam was nearing completion in 1894. Greenwood (1960:103, most likely using McClintock 1921:183 as his source) claims that construction was initiated in 1886. McClintock, however, only states that the St. Johns Irrigation Company was founded in 1886, and does not indicate how long after this date actual construction was begun on the dam.
24. The original Slough Dam was reported to have flooded only 500 acres of land (see Jensen n.d.d.:8/9/1894).
25. Between 1905 and 1910 the Mormon population at St. Johns declined from 566 persons to only 455. This decline did not reverse itself until the activities associated with the building of the subsequent Lyman Dam presented new opportunities.
26. The Denver company sold its interest in the Lyman Reservoir to the people of St. Johns in 1914 (see *Snowflake Herald*:4/16/1915).

27. Great expectations were focused on this joint venture. It was anticipated that the completion of such a large dam and reservoir would permit the intensive cultivation of 15,000 acres of land in the St. Johns area (see Berry 1910:18).

28. The *Snowflake Herald* (4/16/1915) reported that, in addition to its total destruction in 1915, the Lyman Dam, "the constant work of 12 years", was also washed out once during its construction.

The collapse of Lyman Dam in 1915 precipitated the destruction of dams downriver at Meadows, Hunt and Woodruff. While the dam at St. Joseph withstood the flooding, damage to this dam was estimated to be about $1,000 (ibid:4/23/1915). The collapse of the Lyman Dam was felt throughout the lower settlements. The following list of "direct losses" was included in the *Snowflake Herald* (5/7/1915):

Damage to Lyman Dam and canals	$ 90,000
Houses washed out at St. Johns	7,000
St. Johns Irrigation Company	3,000
Bridge at St. Johns	2,500
Bridge at Hunt	2,000
Crops and fences at St. Johns and Meadows	10,000
Meadows Dam	3,000
Udall Dam	18,000
Crops, fences, ditches at Hunt	8,000
Woodruff Dam	17,000
Crops, etc. at Woodruff	13,000
5-mile steel bridge	4,500
Holbrook bridge	5,000
Other (including livestock, etc.)	5,000
Total Direct Losses	$196,000

"Indirect losses" were listed as follows:

Crop loss at St. Johns	$60,000
" " " Meadows	5,000
" " " Hunt	12,000
" " " Woodruff	10,000
Total Indirect Losses	$87,000

Several of these estimates of losses proved too small, most notably the cost of repair to dams at both St. Johns and Woodruff. Moreover, as the article pointed out, "it is safe to say that other indirect losses will make the total up to a full hundred thousand (*ibid.*)." The loss of a dam at St. Johns was clearly a devastation imposed throughout a much larger proportion of the basin.

29. The *St. Johns Herald* (1/30/1919; 3/6/1919; 3/27/1919; 5/15/1919) inicated that the Arizona State Loan Board took over actual construction of the new Lyman Dam and pushed it to completion under the direction of their own engineers. As of May 15, 1919, 125 men and 70 teams were employed on the dam, and more were being sought. Expectations were that the dam would be high enough by June 15th to provide water for the coming season. Apparently, until the state took over, reconstruction of the Lyman

Dam had completely ceased (*ibid.*: 1/30/1919), placing agricultural productivity at St. Johns in a state of uncertainty for three or four years.

30. This figure represents all of the irrigated farmland within the boundaries of the Bureau of Reclamation's Snowflake Project Area, which included "the area adjoining the towns of Snowflake, Shumway and Taylor" (Bureau of Reclamation 1947:1). The project boundaries encompassed a total area of 27,500 acres, with most of the irrigated farmland in this territory located in the immediate Snow-flake-Taylor area.

31. Because the original residents of Taylor had settled in the valley with Stinson's approval prior to the sale of his property to William Flake, many members of this settlement disputed the extent to which succeeding settlers at Snowflake had prior claim to the waters of Silver Creek. This conflict was settled in 1883 when the following arbitrator's report was submitted to the farmers of both towns:

> That in order to make suitable reservoirs and water ditches Snowflake shall pay at the rate of three dollars per acre and Taylor at the rate of five dollars per acre for all their farming land and city lots watered or that may hereafter be watered out of the waters of Silver Creek....(Quoted in Fish n.d.a.:41; also quoted in LeVine 1977:112).

In 1888, the assessment rates were apportioned equally among the farmers of both settlements (*ibid.*), and a formal integration of the irrigation systems of the two towns was achieved in 1893 with the incorporation of the Snowflake and Taylor Irrigation Company (LeVine 1977:113).

32. The expenditures for surveying, legal fees and the purchase of headgates associated with the construction of Daggs Dam totaled $3,800 (LeVine 1977:113). The remaining costs covered the actual construction of the dam.

33. LeVine (1977:155) mentions that in 1916 Daggs Dam, as well as other dams at Snowflake and Taylor, needed repair due to the stresses of high water. The following year the Apache Railroad was granted permission to use the dam as a right of way for its roadbed (*ibid.*), which led to a strengthening of the dam by the railroad company. The Dam needed repair again in 1932, and the Apache Railroad dispatched a crew of 20 men to aid in this project (*ibid.*:116; see also *Snowflake Herald*:11/13/1931).

34. The most serious setback to farmers at Snowflake and Taylor with regard to their investment in an irrigation system was experienced long after the period under consideration here. In September of 1935, work began on the construction of the Lone Pine Dam, which was expected to increase irrigated acreage in the two valleys by about 4,000 acres. Completed in March of 1937 at a total indebtedness of nearly $300,000, this dam stood 95 feet high and was expected to impound 11,000 acre-feet of water (see LeVine 1977:116 for a discussion of the financial information and other specifics on the Lone Pine Dam). However, due to extensive faulting throughout the Snowflake-Taylor area (see Chapter 4, footnote 11), this dam proved incapable of storing water. Even after successive attempts to patch various leaks, the reliable storage capacity of Lone Pine Reservoir remains an insignificant 31 acre-feet. The failure of this dam might have resulted in a widespread loss of land by farmers in the Snowflake-Taylor area had the irrigation company not been successful in transferring its indebteness from the land bonded to finance the project to the power division of the irrigation district (see *Snowflake Herald*:5/10/1940).

35. Greenwood (1960:101) does indicate that the canal and reservoir system presently used by the Round Valley Water Users Association was constructed between 1880 and 1898 under the direction of the Mormon Church, which at various times furnished a subsistence pay of $1.25 per day as well as the necessary dynamite. No mention is made of any major setbacks in this construction process.

36. See, for example, Chapter 2, footnote 32.

Chapter 6

Exploiting Environmental Diversity

The little Colorado River Basin is composed of several widely distributed and structurally distinct habitats which are often differentially affected by the same environmental variation and which offer different productive potentials for agricultural populations. Any attempt by the nineteenth century Mormon settlers to exploit these diverse habitats within a single resource-flow system was likely to enhance local community stability. Stability would be increased by reducing each individual settlement's dependence on a specific habitat and by providing additional resources from other, independent habitats which were unlikely to suffer from the same schedule of variation. Several early Mormon adaptive strategies either manifestly or latently incorporated a multi-habitat adaptation within a single resource-flow system. To the extent that this occurred, specific settlements were more likely to weather potentially devastating environmental fluctuations and less likely to succumb to the stresses that such instabilities imposed.

The success of an adaptive response to environmental conditions depends in part upon the resources that a community is able to direct into appropriate channels. Accordingly, the size of their population, the technology which they commanded, and the organization through which they directed their labor and other resources determined the ability of the Little Colorado Mormon settlements to effectively exploit the opportunities presented by a multi-habitat adaptation. However, simultaneous demands upon a population's resources limited the amount of energy that could be directed into the expansion or integration of resource flows originating in distinct and widely separated habitats. Due to insufficient manpower and to the drains imposed on that manpower by fluctuations in the conditions defining their primary habitat, early settlements in the lower valley of the Little Colorado River were unable to effectively expand their exploitative activities into other, frequently more productive locations. The effective incorporation of diverse habitats into a single resource-flow system was only achieved at a later date when increased immigration and a regionally coherent church organization permitted the synchronization of resource flows from a variety of widely distributed habitats throughout the basin.

UNITED ORDER CONJOINT ENTERPRISES

Although their emphasis from the beginning was to establish viable agricultural settlements based on the flow of the Little Colorado River, early Mormon colonists quickly recognized the adaptive advantage in exploiting habitats away from the river which might supplement their primary agricultural productivity. As Peterson (1967:205) notes,

> economically, the life of all the Little Colorado United Orders was dependent upon an extensive use of surrounding resources rather than an intensive agricultural use of village grounds.

Although the exploitation of diverse habitats was manifestly based on the nineteenth century Mormon desire to remain self-sufficient and separate from the influence and control of the dominant Gentile society,[1] it also represented an appropriate adaptive strategy for these early settlements given the unreliability of farming productivity in the lower Little Colorado River Valley. Not only did those enterprises established to exploit specific resources away from the Little Colorado River furnished their provisions at times when, due largely to dam failures, agricultural productivity along the river was a complete failure, but the per capita productivity of these supplementary operations was frequently greater and more stable than that achieved through irrigated farming in the lower valley.

An appreciation of the complete costs imposed by recurring dam failures among lower valley settlements must include a consideration of the opportunities lost by these towns to exploit other more productive habitats in the basin which might have provided a buffer compensating for the insufficiencies of their farming productivity. The continually high labor cost of farming within the lower valley, combined with the inability to attract and retain a sufficient number of settlers there, precluded the exploitation by these settlements of supplementary resources far removed from the river valley. Caught between the labor needed in farming (especially in rebuilding dams) and the continuing exodus of manpower, the systematic exploitation of nonfarming activities gradually declined until they were completely abandoned.

The Sawmill

The four original United Order settlements conjointly operated four enterprises: a sawmill, a dairy, a gristmill and a tannery.[2] The first of these enterprises was the sawmill. Donated to the newly founded colonies by the church in Salt Lake City, a steam sawmill was erected during the fall of 1876 and put into operation on November 6th of that year (Porter n.d.a:65). Located about 60 miles southwest of Sunset, production at the sawmill between November and the end of 1876 was 51,202 board-feet at a cost of 1199.75 man-days of required labor (*ibid.*).[3] By April of the following year, the sawmill produced 100,000 board-feet of lumber (Jensen n.d.a.:4/10/1877), and during the month of November, 1878 an additional 80,000 board-feet of lumber was sawed (Little Colorado Stake:11/30/1878).

The sawmill supplied lumber for 3 years. During that time, most of the lumber produced was divided among the settlements operating the mill, as a large quantity of building materials was initially needed to establish these colonies. In 1879, however, 39,473 board-feet was charged to individuals at Snowflake (Porter

n.d.a.:65), and as early as September, 1877 a delegation was sent to Prescott, Arizona in order to determine the current market price that might be obtained for any lumber sold (St. Joseph United Order:9/6/1877).
Maintaining a sufficient supply of labor at the sawmill soon proved to be a problem for the early settlements. Because the long haul by team from the mill to the settlements took most of a week round trip (Tanner and Richards 1977:34), a more or less permanent population had to be retained there. However, after the heavy initial input of labor, the number of permanent residents at the mill steadily declined, and by August of 1878 it was reported that "there are only part of three families left" (Little Colorado Stake:9/3/1878). Successive reports in the same minutes dated November 4, 1878 and March 7, 1879 reveal difficulty in retaining sufficient numbers at the sawmill location and, therefore, of being unable "to keep school longer than three weeks" (*ibid.*:3/7/1879). During 1881, a report indicated that "at Millville at present there is but one family --hence no meetings and schools" (*ibid.*:11/2/1881), and by the close of that year the mill was sold to William Flake at Snowflake.

The Mormon Dairy

Perhaps the most successful of the abbreviated conjoint enterprises was the dairy. Established in 1878, the dairy provided substantial and essential provisions for the struggling colonies, as Tanner's (Little Colorado Stake:216-B)[4] comment suggests.

> The dairy at Pleasant Valley which came to be known as the Mormon Dairy was probably the brightest spot in the enterprise of the Little Colorado settlers. In addition to cheese they made butter and raised hogs. They were also successful in raising better potatoes than were produced on the river.

In 1878, 48 men and 41 women from Sunset and Brigham City (the two largest settlements) were at the dairy caring for 115 cows and making butter and cheese (McClintock 1921:154). By the end of August of that year the dairy "had on hand 4,000 lbs. of cheese made since the first of July" (Little Colorado Stake:8/31/1878), and by November of the same year Nuttall during his trip to the region reported that the dairy had produced a total of 5,400 pounds of cheese and 442 pounds of butter (see Peterson 1973:111). Near the close of 1878 it was reported that 5 families had cared for 115 cows since June 1st of that year and that some 75 head were in actual use (see Jensen n.d.a:11/4/1878). These same five families erected 3 dwelling houses (16 feet by 18 feet), a dining hall (32 feet by 16 feet), as well as corrals, pens and other structures needed to care for the livestock (*ibid.*).

While no figures are available for either 1879 or 1880, one report for 1880 claims that "the dairy interest had prospered this year. Stock were doing well. Tannery was doing well considering our inexperience in that line" (Little Colorado Stake:11/2/1880). Production at the dairy for 1881 was recorded as follows: 2,000 pounds of cheese, 1,300 pounds of butter, 340 bushels of potatoes and 1,100 pounds of pork (Warner 1968:10). By 1882, 200 cows were reported being milked at the dairy (*ibid.*:11).

As population size at both Sunset and Brigham City declined, the Mormon Dairy also suffered. With the abandonment of Brigham City in 1881, its interest in the dairy was purchased by the church and transferred to the remaining settlements. This was done to support the Little Colorado Stake tithing heard,[5] and after 1881 large numbers of cattle belonging to the Little Colorado Stake were being maintained at Pleasant Valley. One report (see Smith 1934:8) listed 500 head of cattle for Sunset, 300 head for Brigham City and 300 head for St. Joseph, in addition to about 200 horses.[6] In 1882 the church cattle were moved from Pleasant Valley to Wilford (near Heber), thus founding that place (Porter n.d.a.:79).

Although the dairy was successful and enhanced the aggregate productivity of the lower valley settlements, its productivity ultimately depended on the availability of surplus labor from these same towns. Moreover, because it was located at a considerable distance from the river valley, the efficient operation of the dairy, like the sawmill, demanded a resident population. As the parent settlements endured recurring hardships trying to direct sufficient labor into maintaining their fragile irrigation systems, support for the dairy became increasingly difficult--eventually impossible--to provide.

Labor at both the sawmill and the dairy was most acutely needed during the summer and fall. The weather at higher elevations during the winter, including substantial accumulation of snow, made logging activities difficult, if not impossible, to perform at this time. Similarly, weather and seasonal variation in range conditions made dairying a largely summer activity at higher elevations. Consequently, while men and women from the various colonies were assigned to spend a summer at either the sawmill or the dairy only a skeleton population was retained at either location during the winter, primarily to maintain buildings and other structures. Consequently, any labor performed at the sawmill or the dairy conflicted directly with that needed to sustain farming along the Little Colorado.

Porter (n.d.a.:375) provides a record of labor expended by the settlers at St. Joseph during 1879 (see Table 6.1). Except for 92.5 days ditchwork (cooking for men working on the ditch), 125 days at the dairy, and 37 days teaching school which were attributed to women, the remainder of this labor (4,506.5 man days) was credited to 23 men, producing an average of 196 man-days labor per man for that year.[8] While certain nonseasonal labor (e.g., some smith work, carpentry work, shoemaking, and building and sundry labors) could be scheduled for the less congested winter months, much of the remaining work had to be performed concurrently during the summer. Having to complete the bulk of the labor expended at the dairy and the sawmill at same time that work had to be invested in the farm, ditches and other agricultural activities (in addition to those routine functions required to maintain the community) placed a considerable burden on the limited labor force available.[9] During periods of critical labor shortage, which occurred nearly every year due to recurring dam failures, the stress engendered by investing in supplementary productive activities became particularly acute and increasingly difficult for the settlements along the lower valley to endure. Thus, unstable and insufficient levels of productivity in the primary sector (farming) precluded the continued investment of resources into potentially productive supplementary resource flows.

Table 6.1

Labor Expended at St. Joseph
1879

Work Performed	Man-Days Labor Expended
ditch	764.5
farm	734
garden	154
stock	256.5
herding cows	90
herding sheep and shearing	326
dairying	473
freighting to Utah	128
choring	234
carpenter work	303.75
shoemaking and shoemaking work	82
smith work	95
building and sundry labor	710.25
teaching school	95
sawmill	315
Total[7]	4,761

SOURCE: Porter (n.d.a.:375).

Labor Shortages

Created by repeated dam failures and the inability to retain sufficient immigrants, the endemic labor shortage which characterized the lower valley settlements was exacerbated by the relatively high dependency ratios associated with these towns. Small to begin with, populations in the lower valley settlements maintained ratios of consumers to producers that reduced their productive capacity below that already suggested by their limited size.

In 1877, for example, children under 8 years of age comprised 35.5% of the population of St. Joseph.[10] The proportion of the population under 8 years of age there increased to 43% in 1879 and to 49.5% in 1881, a year which witnessed a major dam reconstruction. Children under eight still accounted for 42.5% of the population of St. Joseph in 1884, another year of a major dam undertaking. The proportion of children under eight among other settlements in the lower valley were comparable to St. Joseph (see Table 6.2). In 1883 Sunset, with a population of only 63 persons, achieved the peak dependency ratio in the lower valley. Over 50% of its population was less than 8 years of age, providing an inadequate labor force to sustain agricultural activities at this settlement. This was the year that Sunset was abandoned.

The proportion of the population under 8 years of age actually increased during the early years of settlement. While the initial population founding St. Joseph was listed as 45 men, 13 women, 4 boys and 11 children, making for a 20.5% dependent population (15/73), the average proportion of the population under 8 years of age between 1877 and 1880 was 40.4% and increased to 45.4% during 1881 to 1884. This same general increase is reflected in the aggregate figures for the lower valley as a whole. In 1877, 33.3% of the combined lower valley population was under 8 years of age, which increased to 36.8% in 1878 and to 46.6% in 1884.

Children under 8 years of age do not represent the total dependent segment of a population. Although elderly persons were conspicuously absent among early pioneer populations in the region, many children existed in these populations which were over eight years of age.[11] The size of this latter cohort is difficult to determine. The letters which John Nuttall sent to Utah during his tour of the Little Colorado settlements provide some clues, however. Nuttall (1878a, 1878b, 1878c) listed the number of men, women, boys and girls present in each of the settlements that he visited (see Table 6.3). Considering the categories boys and girls as defining the dependent population, the dependency ratios for these settlements increase dramatically over those calculated simply on the basis of children under 8. While the settlers at St. Joseph registered 45% of their population as under eight years of age, fully 60% of that population were classified as children. Likewise, while Sunset recorded 21% of its population as under 8 years of age in 1878, nearly 64% were classified as children that year. Equally marked differences between these two figures existed among the other lower valley settlements.

Table 6.2

Percent of the Population under 8 Years of Age among Settlements
in the Lower Valley of the Little Colorado River
1877 - 1886

Year	Sunset	Brigham City	St. Joseph
1877	25.0	34.0	36.5
1878	21.0	36.0	45.0
1879	33.8	40.2	43.0
1880	31.6	38.7	38.8
1881	34.5	--	49.5
1882	43.6	--	45.8
1883	52.4	--	43.6
1884	41.5	--	42.5
1885	42.3	--	23.3
1886	--	--	23.1

SOURCE: Little Colorado Stake (n.d.).

Although comparable data on the number of children under 8 was not obtained for settlements outside the lower river valley, the available data suggests that the proportion of children to the total population was considerably lower in those towns away from the lower river, particularly Snowflake. The number of children listed as under 8 years of age for the Eastern Arizona Stake at the close of 1879 was only 30.9% of the total stake population. Although the proportion of this cohort increased somewhat during subsequent years, equaling 36.7% in 1883 and 35.8% in 1884, it was still considerably less than the proportion displayed by the lower valley settlements. Moreover, in 1880 children under 8 years of age comprised only 26.2% of the population of Snowflake. In addition, a comparison of the mean family size (suggesting the relative proportion of dependents) at Snowflake and St. Joseph during the 1890's reveals a consistent and marked difference between these two towns (see Table 6.4).

As a result of the high dependency ratios that prevailed among the lower valley settlements, colonists in this portion of the basin suffered an additional labor shortage to that already imposed on them by their small numbers.

Consequently, during the critical years when these early settlers struggled to establish the productive bases of their communities, the reduced manpower implied by such high proportions of children in the population seriously compromised their ability to direct sufficient energy into channels capable of offsetting environmental instability. The lower valley settlements never seemed to command enough resources to repair dams, clean ditches or prepare fields, let alone expand into new habitats which might prove potentially more productive and reliable than farming alkaline soils by means of saline and silt-laden irrigation water obtained from an unpredictable river.

Table 6.3

Men, Women, Boys and Girls in the Lower Valley Settlements
September 1878

Settlement	Men	Women	Boys	Girls	Total	Percent Boys and Girls
Sunset	19	18	43	22	102	63.7
Brigham City	43	46	61	60	210	57.6
St. Joseph	13	14	22	18	67	59.7
Woodruff	15	15	25	20	75	60.0
Lower Valley	90	93	151	120	454	59.7

SOURCE: Nuttall (1878a, 1878b, 1878c).

TITHING REDISTRIBUTION

Following the failure of the conjoint enterprises, Mormon pioneers in the Little Colorado River Basin applied several techniques that resulted in the effective redistribution of tithing surpluses among the numerous, widely dispersed settlements throughout the region.[12] The effectiveness of tithing redistribution depended on several supporting institutions, including: (1) a regional board of trade that established uniform prices among Mormon settlements in the region, (2) a network of church-affiliated mercantile enterprises that integrated local Mormon settlements into a regionally coherent economic system, and (3) a system of quarterly stake conferences at which representatives from each settlement exchanged information, assessed local needs and determined appropriate resource allocations. Following the lead established by parent institutions in Salt Lake City, each of these enterprises represented a conscious attempt by local church leaders to extend the cooperative thrust of the languishing United Order Movement in the face of new realities. Etically (see Harris 1968:575), these interlocking operations provided a redundancy in the flow of resources that assured the availability of a minimum supply of resources among the various Mormon settlements in the basin.

Tithing

Tithing occupies a central place in the communal emphasis of the Mormon religion, as well as in the financial support of the Mormon Church. Ideally, all Mormons are expected to tithe ten percent of their annual income to help finance the many and varied activities directed by the church. During the nineteenth century these included immigration, colonization, education, welfare, the construction of such major public works such as factories, irrigations systems, railroads and temples, and the operation of numerous mercantile establishments (see Arrington 1958:*passim*). Representing a form of "divine taxation", tithing is collected by the local wards, centralized into each stake office, and then forwarded (less a portion granted for local use) to the General Tithing Office in Salt Lake City.

Throughout the nineteenth century and into the early years of the twentieth century, nearly all of the tithing collected among the Little Colorado Mormon settlements was paid in kind, as cash was a scarce commodity in this frontier region (cf. Peterson 1973:246-247; Leone 1979:56-62, 229-241). Objects were tithed as they became available, and tithing stocks were generally most abundant during the fall following the harvest. Because tithing stocks represented surplus productivity, both aggregate tithing and the tithing of specific items depended ultimately on the total productivity registered in the respective categories during any particular year. Due to the existence of marked spatial and temporal differences in items produced within the basin, substantial variation occurred in the amount and kind of tithing collected within and between specific settlements.[13]

Only a small fraction of the tithing gathered was actually forwarded to Salt Lake City during the nineteenth century. Leone (1979:56) estimates that between 75% and 90% of the tithing collected remained within the Little Colorado River Basin. A substantial surplus was, therefore, available to subsidize specific projects deemed important by local church leaders who were granted considerable discretion by church authorities in Salt Lake City regarding their allocation of tithing resources. Tithing, thus, functioned to store the surplus produced at one spatio-temporal locus in order to offset production deficiencies at different times and/or places. Due to the vagaries of the natural environment in this basin, differential productivity was likely to occur: (1) between individuals within the same town, (2) within the same settlement during different years, and (3) at different settlements during the same year. Tithing stocks were also exchanged between towns which suffered deficiencies of different products in order to offset local shortages of specific commodities. Leone (1979:74) indicates that up to 25% of the total annual tithing collected among the Little Colorado settlements during the nineteenth century was exchanged for such "better kinds".

Table 6.4

Mean Family Size at St. Joseph and Snowflake
1892 - 1897

Year	St. Joseph			Snowflake		
	Population	Number of Families	Mean Family Size	Population	Number of Families	Mean Family Size
1892	117	16	7.3	450	74	6.1
1893	107	15	7.0	446	81	5.5
1894	116	16	7.3	463	81	5.7
1895	128	17	7.5	399	67	6.0
1896	133	18	7.4	407	70	5.8
1897	132	18	7.3	421	71	5.9

SOURCE: Historical Department, Church of Jesus Christ of Latter-Day Saints (n.d.).

Clearly, the most important function performed by tithing redistribution from the perspective of establishing viable agricultural communities in the basin was in subsidizing those strategically important projects that were essential to the survival of individual settlements. Towards this end, the consolidation of tithing resources into a central stake reserve significantly enhanced the effectiveness of tithing redistribution. The chances for success of numerous projects, many of which were substantial in scope and thus beyond the limited resources of individual settlements, were generally increased (and frequently assured) by the clear channels of resource allocation provided by the regional church organization. As a mechanism for counteracting the impact of uncertain and highly variable environmental conditions, tithing collection and redistribution functioned to disperse the drains and consolidate the subsidies accompanying such resource flows. In the absence of an effective secular administrative apparatus, the necessary investments in "public works" (infrastructure) in the basin were made possible by the redistributive flow of resources through an equally-centralized church organization.

As already discussed (see Chapter 5), the single most important public works among the Little Colorado Mormon settlements were the vital irrigation systems upon which agriculture in this arid basin depended. The destruction of these systems threatened an entire community, making their immediate reconstruction essential. Because such undertakings were frequently beyond the resources of individual settlements, most notably the smaller ones, it was in subsidizing dam reconstructions that tithing redistribution performed its most effective adaptive function.[14] By furnishing threatened settlements with the supplies needed to sustain them until the next harvest; by allocating tithing funds to defray the cost of hiring labor to assist in dam reconstruction; and by allowing individuals to offset their tithing obligations by donating labor on church-approved projects, local church leaders were able to channel considerable surplus resources into those activities which were of strategic importance to community maintenance and, thus, the success of the colonization effort.[15] It is doubtful whether either St. Joseph or Woodruff would have withstood the successive destruction of so many dams had it not been for the numerous and substantial energy subsidies they received as a result of tithing redistribution. Tithing redistribution was, therefore, largely responsible for the survival of several settlements and, thus, performed a critical role in successful Mormon colonization of the basin. However, the effectiveness of tithing redistribution was largely dependent on the existence of another Mormon institution established within the basin, the Board of Trade.

The Board of Trade

Responding to the decline of United Order settlements throughout its domain, Church authorities established Zion's Central Board of Trade in 1878 to sustain their quest for Mormon self-sufficiency (see Arrington 1958:341ff.; Peterson 1973:124ff.). Designed to encourage and direct cooperative production, marketing and purchasing ventures, the Central Board of Trade also regulated wages and prices accepted in economic transactions with non-Mormon business interests. By controlling the economic exchange with Gentile traders, Church leaders hoped to eliminate the destructive price competition that existed among Mormon producers.[16] By establishing specific rates charged to outsiders, the Board of Trade, in effect, fixed the prices that prevailed among Mormons as well.

Pursuant to the integrative purpose of the Board of Trade Movement, local boards of trade were founded in stakes throughout Zion which were administratively united with the Central Board in Salt Lake City. Aspirations for self-sufficiency were, thus, more realistically transferred from individual towns to more encompassing regions and territories and, ultimately, to an economically coherent total Mormon community. Annual meetings were held which brought together representatives of the Central Board and those of the numerous local and regional boards. With input accepted from local and regional affiliates, the Church through the Central Board of Trade was able to effectively channel its considerable resources into those projects throughout its far-flung satellite regions that served to "look after the manufacturing, mercantile and other interests of Zion" (see Arrington 1958:344).[17]

In conjunction with the larger church movement and in the wake of declining United Order settlements locally, stake boards of trade were established within the Little Colorado River Basin which regulated Mormon productive and distributive activities in this region. With the same primary goal of eliminating pricing competition among the Little Colorado Mormon towns which could be exploited by Gentile traders,[18] uniform prices were established for Mormon produce throughout the basin.[19]

Because prices established by the Board of Trade ultimately prevailed in exchanges that took place among Mormons as well, pricing policies established in the Little Colorado River Basin facilitated the redistribution of resources among the various Mormon settlements and inhibited the expatriation of these resources from the redistribution system. By establishing set prices for all commodities, items in particular abundance at one town might be converted more profitably through exchange within the Church's local marketing apparatus--either donated as tithing or deposited for credit in the largely Church-run cooperative stores (see below). Since both tithing stocks and a cooperative store's inventory largely reflected local surpluses, stake offices as well as individuals could exchange surplus goods on hand or those which were scarce locally (provided they were abundant elsewhere) and at prices that were not harshly responsive to market fluctuations. As already indicated, a substantial amount of such "trading-up" occurred.

The ACMI

The quest for self-sufficiency among the Little Colorado Mormon settlements culminated in the formation of the Arizona Cooperative Mercantile Institution (ACMI) (for other discussions of the ACMI see Peterson 1973:136-153; Leone

1979:79-82). As with the Board of Trade to which it was closely linked, the ACMI developed as a local component of an encompassing church institution--the Zion's Cooperative Mercantile Institution (ZCMI). Although the ZCMI was originally established by the church in Salt Lake City in 1868 to consolidate and monopolize Mormon trade and, thus, reduce Gentile expropriation of the capital and wealth of Zion, this enterprise quickly evolved into a complex and widely distributed manufacturing, trading and banking institution operating throughout the Mormon domain (see Arrington 1958:298-322 for a discussion of the ZCMI and its varied operations). A similar evolution, albeit less elaborate and on a considerably smaller scale, occurred within the ACMI.

The ACMI was founded in 1881 during the June conference of the Eastern Arizona Stake, and 11 local church leaders comprised its initial board of directors.[20] Begun with a subscription of $1,200 (Fish, n.d.a.:9), capital stock in the institution increased to $5,000 by that October (*ibid.*:10) and to nearly $20,000 by the close of 1882 (*ibid.*). The central ACMI Store was originally located in Holbrook, but was quickly moved to Woodruff where it remained for several years.[21] Coincident with the founding of the central ACMI, locally-owned affiliates were established in nearly every Mormon settlement in the region, and total sales by the ACMI increased from $65,000 in 1881 to $85,000 in 1883 and to $111,000 in 1891 (see Fish n.d.a.:9-10, n.d.b.:265, 302; Peterson 973:140).

While the AMCI did not shun business with Gentiles--indeed, such transactions represented a not inconsiderable component of the store's retail trade (cf. Peterson 1973:140)--its wholesale operations were undertaken almost exclusively as an agent for the many consociated cooperative establishments within the basin. Purchasing at prices that individual stores could not command, the ACMI was able to import needed materials at what were likely the lowest prices obtainable in this frontier region.

Perhaps the most important functions of the ACMI involved the credit and redistributive roles that this establishment performed. Trade in barter was a central feature of the exchange that occurred through the ACMI. Individuals and affiliated co-ops were able to obtain advances from stocks on hand which could be repaid later, usually after the coming harvest. With liberal credit arrangements, the ACMI facilitated recurring investments in productivity, as it normally carried a heavy load of creditors from one year to the next.[22]

The ACMI also functioned as an important channel for tithing redistribution. Local bishops could exchange goods from their tithing stocks for cash, and often stored tithing produce on credit with the ACMI. Individuals could deposit their tithing obligations to the church at the local cooperative store and have the proper amount credited either to their tithing account or against future needs. By these various credit arrangements and through its intimate association with the local Church leadership, the ACMI transformed an otherwise dormant surplus into a flow of resources that quickly circulated between locations of relative abundance and those of critical need.[23]

Labor was also an important commodity exchanged for goods at the ACMI. Indeed, Peterson (1973:146) maintains that "the very foundations of the ACMI rested upon the barter of time for stock in the institution." A not inconsequential number of persons temporarily obtained a livelihood by freighting[24] for the company or by performing other duties necessary to its maintenance.[25]

The dispersed ownership of the ACMI did not endure for very long. Within two years, possession of the institution passed into the hands of a very few persons, generally leaders in the local Church organization (see Peterson 1973:151), and by the turn of the century none of the local co-ops, save perhaps that at St. Johns, remained under independent management (ibid.:149-150). During the early years of the ACMI when little competition existed, stores were able to pay substantial dividends. In the first year of its operation (1881) the St. Johns co-op store declared a dividend of 25%, while the Snowflake co-op paid between 30% and 35% dividends each year prior to 1885 (ibid.:152). However, with increasing competition, losses due to withdrawals of interest in the company during the polygamy raids of the 1880's (see Chapter 7), to overextensions of credit, and to the drought and panic of the 1890's resulted in sharply reduced dividends. During 1888 and 1889 the ACMI declared a dividend of 20%, and by 1890 this figure had slipped to only 15% (LeVine 1977:38). In 1892 and 1893 the dividends declined further to 6% and 7.5% respectively (Peterson 1973:152). Owing primarily to overextensions of credit, many of the more marginal stores (particularly those in the poorer mountain settlements) either closed their doors or were sold to the parent establishment.[26]

The dispersed ownership of the ACMI was lost rather early in its existence. Moreover this institution never quite lived up to the cooperative claims of its founders (see Peterson 1973:151-152). However, due to its inextricable ties to the local church, the ACMI through its varied credit and trading activities provided an effective, centrally directed redistributive mechanism capable of offsetting the instabilities encountered by Mormon towns in the basin during the settlement period.

Quarterly Stake Conferences

Another essential component of the tithing redistribution system was the cycle of quarterly stake conferences attended by representatives from each ward within the stake. Held ostensibly for religious communion, these conferences were the focus of a substantial information exchange regarding the material conditions present at individual settlements (see Little Colorado Stake n.d.; St. Johns Stake n.d.; Snowflake Stake n.d.). During every conference a representative (usually the bishop) from each ward presented a report which outlined the prevailing circumstances at their settlement, and much discussion (and lobbying) took place regarding the judicious allocation of tithing funds. The quarterly occurrence of these conferences--before the planting season (February-March), after the spring runoff (May-June), following the intense summer rains (August-September) and subsequent to the fall harvest (November-December)--assured a continuous appraisal of local material conditions obtaining throughout the basin. Thus, during critical junctures in both the agricultural and seasonal cycles, information concerning the impact of the latter system on the former could be rapidly transferred to the larger community and a solution to acute problems quickly executed.

CONCLUSION

Several adaptive strategies were employed by nineteenth century Mormon settlers in the Little Colorado River Basin which integrated the exploitation of several distinct habitats into a single resource-flow system. Two general systems stand out: (1) the conjoint enterprises of the early United Order settlements in the lower valley, and (2) tithing redistribution among the numerous settlements scattered throughout the river basin.

Although the four conjoint enterprises established by the United Order settlements were manifestly instituted to offset environmental variability, these early attempts at multi-habitat exploitation were not ecologically viable given the material conditions encountered by populations inhabiting the lower valley. With their effective manpower reduced by burdensome dependency ratios, these small populations suffered heavy drains on their meager resources due to the variability and power of the Little Colorado River. Chronic labor shortages and inadequate harvests, aggravated by recurring dam failures, precluded the investment by these settlements of sufficient resources to effectively exploit habitats geographically removed from the Little Colorado River Valley. The conjoint enterprises, thus, proved ineffective as mechanisms of environmental regulation despite their manifest ethnoecological basis and the cooperative Mormon values upon which they were based.

Tithing redistribution, on the other hand, succeeded as a mechanism of environmental regulation, despite the fact that its critical supporting institutions were instituted for manifestly nonecological purposes. Although tithing redistribution itself was explicitly applied to offset the effects of environmental variability, the Board of Trade and the ACMI were established in response to competing non-Mormon business interests, while the quarterly stake conferences functioned ostensibly to promote religious unity. However, despite their non-ecological intent, these latter three institutions provided the framework for a more ecologically viable multihabitat resource-flow system than had been attempted by the early United Order settlements. Moreover, the system of tithing redistribution succeeded for the same reasons that the conjoint enterprises failed. By broadening the scope of the exchange network (organizationally and spatially) and by incorporating within this network a sufficiently large aggregate population exploiting numerous, independent, and widely dispersed habitats, this subsequent, more inclusive resource-flow system was more suitably structured to counteract environmental instability within the basin.

By including every Mormon settlement in the basin and, thus, integrating resource flows from every exploited habitat, the later regionally-coherent exchange system was able to generate quite substantial resource flows at those very times when labor and other resources were critically needed to offset the destabilizing impact of local environmental perturbations. Uniting the many, dispersed Mormon settlements within the jurisdiction of a centrally administered, eco-politico-religious organization[27] served to enhance the responsiveness, redundancy and reliability of tithing redistribution and, therefore, its effectiveness as a buffer to environmental variability. In contrast to the system of conjoint enterprises, which integrated the material resources of merely a few hundred persons inhabiting three or four settlements in similar highly unstable primary habitats, the system of tithing redistribution integrated the productivity and labor of over 2,000 individuals inhabiting nearly two dozen settlements and exploiting numerous, widely separated habitats which were unlikely to suffer the same schedule of environmental variation. Consequently, tithing redistribution provided greater redundancy in the flow of resources. It was, therefore, better able to offset the drains imposed by local habitat instability and, thus, function as a mechanisms of environmental regulation

An important factor contributing to the effectiveness of tithing redistribution was the greater region-wide access that this system provided to the surplus labor and productivity of the larger, more productive and more stable settlements situated at intermediate elevations. While population size and productivity among the smaller settlements at both higher and lower elevations waxed and waned more or less regularly, those towns located at intermediate elevations--most notably Snowflake, Taylor and St. Johns--repeatedly furnished surpluses that on several occasions were critical to the survival of the smaller, less stable settlements. However, due to the scope and pervasiveness of environmental variation in the region, even the more prosperous intermediate settlements benefitted from tithing redistribution. Between 1887-1905, agricultural productivity varied over 35% above and below its mean at Eagar and 28% above and 40% below its mean at St. Johns. Even Snowflake, the most prosperous settlement in the basin, experienced 30% variation in agricultural productivity during the nineteenth century.

From the perspective of assuring the redistribution of surplus resources, it is also significant that tithing redistribution was made possible by institutional arrangements which were not economically disadvantageous to the donor populations. That the more successful farmers in the more prosperous, intermediate settlements benefitted from the system is suggested by several facts: (1) the larger resource-flow system was principally a product of the Eastern Arizona Stake; (2) the most successful of the cooperative branch stores were located at Snowflake and St. Johns; and (3) these more successful cooperative stores were able to pay their investors substantial dividends.

The success of tithing redistribution as a mechanism of environmental regulation was also due, in part, to its integration with encompassing parent institutions outside the basin (see also Lightfoot 1980). By furnishing resource flows which were not constrained by material conditions within the basin, the enveloping church organization and its various institutions (with their substantial potential resources) frequently able to provide a critical subsidy to offset the destruction wrought by local environmental forces. The availability of church subsidies further increased the redundancy of the system.

The hierarchical, integrated and mutualistic structure of nineteenth century Mormon institutions, thus, provided an adaptive organizational apparatus from which a suitably complex resource-flow system could emerge among the numerous Mormon agricultural settlements scattered throughout the basin. The forced savings of surpluses, the central redistribution of these surpluses, and the repeated availability of material subsidies from Mormon sources external to the basin underlay the adaptive advantage of nineteenth century Mormon institutions under variable, frontier conditions. It is largely for these reasons that the establishment of viable farming settlements throughout the Little Colorado River Basin--and most of the Mountain West--was largely a Mormon achievement.

Not all the environmentally imposed instabilities that affected early Mormon settlers in the Little Colorado River Basin were natural in origin. Because these early settlements existed on the frontier of the larger U.S. political economy, several distinct though interrelated impacts were thrust upon them as a result of the vigorous expansion of that frontier during the late nineteenth century. The following chapter examines the principal components of the expanding U.S. frontier that affected Mormon settlements in the region. The expansion of the frontier into the Little Colorado River Basin affected specific settlements differently and again illustrates the adaptive value of Mormon institutions in the face of environmental stress.[28]

NOTES

1. The exclusion of Gentile immigration and access to vital agricultural resources underlay the Mormon colonization scheme, generally and in the Little Colorado River Basin. This exclusionist policy lay behind the willingness to purchase the "squatter's claims" of Stinson and Barth, even though the prices demanded were considered exorbitant at the time.
2. While information on the sawmill and dairy are scanty, data concerning the gristmill and tannery are almost non-existent. Consequently, the following discussion will only focus on the former two enterprises. These undertakings best represent the early settlers' efforts to exploit the habitat diversity present within the basin and clearly illustrate the problems of resource allocations experience by the lower settlements.
3. Porter (n.d.a.:65) itemizes the labor expended and lumber received by the individual settlements as follows:

Settlement:	Labor Expended (man-days)	Lumber Received (board-feet)
St. Joseph	617	9,682
Brigham City	117.5	10,223
Sunset	310.5	22,916
Obed	154.75	8,381

4. George Tanner recorded numerous comments in footnotes to the Minutes of the Little Colorado Stake. Tanner's comments are included at the end of the typed copy which he prepared and distributed to various archival collections in the region.
5. Due to the scarcity of cash in the Little Colorado River Basin during the nineteenth century, most tithing was paid in kind (see below). Cattle comprised at least a semi-permanent medium of exchange, and substantial stake herds were generally maintained.
6. The importance of livestock operations among the early Little Colorado settlements was considerable and comprised perhaps the most successful aspect of their venture (see Peterson 1970:406). The committee established in August of 1886 to settle the affairs at Sunset after its dissolution reported the following livestock belonging to that company along with the values set by them (Committee of Settlement of the Affairs of the Sunset United Order 1886:11):

525 cows	$25	$13,125
300 calves	6	1,800
250 yearlings	13	3,250
75 steers, 2 years old	20	1,500
80 steers, 3 years old	27	2,160
140 steers, 3 years and over	34	4,761
40 bulls	25	1,000
Total		$27,596

As a Mormon farming village, Sunset was a failure--at least after its initial years when it made a vital contribution.... But as a business it did not fail; it was economically the most notable Mormon success on the Little Colorado. It was never a balanced productive unit with mills, tanneries and textile production providing the good life for its communitants, but as a livestock operation it flourished.... Northern Arizona in the 1880's was a typical livestock frontier. Public domain, the natural reproduction of animals.... (Peterson, 1970:406)

7. Warner (1968:20) claims that a total of 5,669.5 man-days of labor was performed by the men, women and children of the St Joseph United Order in 1879.
8. Eight boys are listed as performing work at St Joseph in 1880, primarily in herding but with some labor performed on the ditch and farm (Porter n.d.a.:375). Information regarding how much labor these boys performed, how many worked in 1879, and whether or not their input is included in the total labor expended is not available.

Exploiting Environmental Diversity 161

9. The representativeness of the labor expended in 1879 for other years cannot be determined. That particular year marked the construction of a dam quite a distance up the river which required considerable ditchwork. However, while the total labor expenditure includes an itemization for the ditch, no mention is made of the dam itself On the other hand, while 734 man-days of labor were performed on the farm in 1879, John Bushman (n.d.) accounted for 1,529.5 man-days rendered there in 1880.

10. One of the categories differentiated in the ward statistical reports of Mormon Church is the number of children under 8 years of age.

11. Although several sources attribute an important economic value to children over 8 years of age in pre-industrial farming systems (cf. Chayanov, 1966; White, 1976; Nag, White and Peet, 1978), the particularly heavy labor and the technological inputs incorporated into agriculture along the Little Colorado depreciated the value of labor performed even by older boys, say under 15 or 16 years of age. This depreciation of youth labor was recognized indigenously through the valuation of adolescent male labor at one-half that of a man, even less than the three-quarter value placed upon female labor Consequently, the allocation of youth labor was generally restricted to selected activities (see footnote 8 in this chapter).

12. The discussion of resource redistribution which follows derives, in part, from that presented by Leone (1972; 1979:Chapter 3).

13. See Tables 2.3, 2.4, 2.5 and 2.7.

14. Both the swiftness required in dam reconstruction and the scale of many of the irrigation systems in relation to the populations affected placed the success of such projects beyond the autonomous capability of most settlements.

15. An integral part of the tithing redistribution system was the role played by the Church through its General Tithing Office. Frequently, the support extended to a local project through the use of stake funds was augmented by even greater subsidies appropriated by church headquarters in Salt Lake City.

16. The Zion's Central Board of Trade was modeled after the Cache Valley Board of Trade. This earlier organization was established in 1872 when 12 cooperatives in that northern Utah valley agreed to act jointly in pricing and in the performance of other marketing functions in order to rid themselves of profiteers who exploited competition among Mormon producers (see Arrington 1958:342).

17. One project which the Central Board of Trade organized and directed was Mormon involvement in the building of the railroad line through the Little Colorado River Basin in 1880 (see Chapter 7).

18. The Minutes of the Little Colorado Stake for August 31, 1879 contain the following entries which illustrate the local concern over pricing competition, as well as the xenophobia directed towards non-Mormons generally.

> Prest. L. Smith...Counseled not to run after the gentiles with our produce or what we have to sell, but let them come to us to buy what they wanted, and all three settlements should have one price and not undersell each other.

> Prest. Jesse N. Smith said he would be willing to cooperate in establishing prices and then we ought to stick to them and not go and under sell (sic.) each other. Spoke of the necessity of letting those outside stores alone. Illustrated the bad results of bringing those who are not of us into our homes.

19. Among the prices established by the local boards of trade within the Little Colorado River Basin were the wages allowed for hand and team labor employed in dam reconstruction.

20. While the initiative to form the ACMI took place during an Eastern Arizona Stake Conference, the store was established as a cooperative enterprise uniting all wards in the Little Colorado River Basin, including those within the Little Colorado Stake. The initial board of directors for the ACMI included the president of each stake (as president and vice-president of the company respectively) and 9 bishops from various wards in both stakes (see Fish n.d.a.:9).

21. Unable to obtain a building site in the new town of Holbrook, the ACMI store was originally relegated to an isolated location along the railroad track. The store subsequently moved up the Little Colorado River to the town of Woodruff where it remained until 1888. During that year the business district of Holbrook burned, and the ACMI was able to acquire a site during the rebuilding of the town.

22. So essential to the farmers in this basin--and so extensively exploited by them--was the credit available through the ACMI that local Church leaders frequently admonished their members for carrying such heavy debts with the cooperative system. Such abuse of credit was proclaimed as an ingredient in the corruption attributed to the larger Gentile society.

23. Because of its ultimate affiliation with the church-wide ZCMI, deposits could even be made at local cooperative stores outside the Little Colorado River Basin, such as in Salt Lake City, and credited to accounts within the ACMI. This credit arrangement facilitated the flow of resources from friends, relatives and other persons and organizations in Utah who were assisting individuals and settlements struggling to establish a foothold in the basin.

24. As indicated in Chapter 5, many of the residents at Woodruff were fundamentally dependent for their existence on employment outside their arduous valley. Many of these individuals worked at least partially as freighters for the ACMI.

25. In addition to the many regular and semi-regular jobs that it provided, the ACMI frequently offered vital employment opportunities for individuals during periods of critical need, thus enabling them to weather difficult circumstances. Such was the case, for example, for the entire settlement of Woodruff following the destruction of their dam in 1884 (see Chapter 5).

26. The ACMI finally closed its doors on March 25, 1933, at which time the Snowflake branch store was re-opened under new management. The Snowflake store operated until the early 1940's when it finally closed for good (see LeVine 1977:38).

27. The diversity of roles performed by the church organization among Mormon settlements in the Little Colorado River Basin has already been discussed. The minutes accompanying the various stake meetings clearly testify to the central involvement of this organization in economic and political matters until the turn of the century. After 1900, a marked decline takes place in open church involvement in secular affairs. This decline is not reversed until the depression of the 1930's, when the Church again assumes a greater role in temporal concerns.

28. For many scholars, Mormon colonization of the Little Colorado River Basin constitutes part of the expansion of the American Frontier. From the perspective of this

monograph, the expansion of the American frontier refers to the westward migration of non-Mormon populations which represented an external impact on this local settlement process.

CHAPTER 7

External Impacts on the Settlement Process

Evolving during the height of successive communal movements within the larger church body, Mormon settlement of the Little Colorado River Basin was initiated and sustained by the desire to establish self-sufficient agricultural communities which were economically, politically and socially aloof from the non-Mormon influences around them. Achieving complete functional integrity served as a principal motivation behind the frequently urgent and concentrated Mormon effort to obtain whatever land claims became available within the basin--often with a considerable devaluation of the costs involved.[1]

A remarkable degree of local autonomy was, indeed, achieved by the Little Colorado Mormon settlements during the nineteenth century--if not individually, then at least as a regional economic system. However, despite the considerable regional unity and exclusiveness achieved by these towns, they were always affected by concurrent material forces originating outside their integrated resource-flow system. From its beginning, Mormon settlement in the Little Colorado River Basin comprised but a small part of a much larger church colonizing effort. Not only was settlement along the Little Colorado conceived, planned and executed under the direction of church leaders in Salt Lake City, but continuous subsidies of money, materials, population (labor), organization and skills were judiciously provided by the church throughout the nineteenth century. As has already been amply illustrated, on numerous occasions the subsidies obtained through their connection with the church prevented the maintenance costs at specific settlements from completely depleting the resources of local populations.

Aside from the church and its considerable resources, several other non-indigenous sources of income and supplies were essential to the survival of early Mormon settlements in the basin. Freighting[2] furnished an important source of income for settlers at Woodruff which offset the repeated failure of farming in that valley. Early attempts to assess the lumber market at Prescott, the formation of a board of trade to eliminate marketing competition among Mormon producers, and the success of the ACMI in its role as an import-export agent also demonstrates the intimate dependence of these early settlements on non-Mormon populations and resources.

What limited degree of isolation may have existed initially was completely eliminated with the coming of the railroad and the subsequent incorporation of the Little Colorado River Basin within the periphery of the expanding American frontier. The railroad's arrival had several important effects on the development of early Mormon settlements in this region. It immediately created markets for Mormon produce and labor, both locally and outside the basin--some of which proved critical to specific settlements. The arrival of the railroad also introduced new burdens, ultimately involving threatened land titles, range competition and deterioration, violence, and political persecution. As with most other impacts, the arrival of the railroad affected individual settlements differently, and church assistance proved to be a critical factor subsidizing community survival.

The Arrival of the Railroad

In 1866, Congress authorized the construction of a railroad line along the 35th parallel. During the spring of 1880, the Atlantic and Pacific Railroad Company began construction along this line in western New Mexico, and within 18 months had pushed 565 miles westward from Isleta to the Colorado River (Peterson 1973:125; see Map 7.1). The timing of the railroad construction was especially beneficial to Mormon pioneers in the Little Colorado River Basin. Following the devastating winter of 1879-1880 (see Chapter 2) and coincident with the general failure of the 1880 harvest, the economic opportunity provided by railroad construction was viewed as a "Godsend" by many local residents (Peterson 1973:125). Income received in return for work on the railroad enabled the purchase of much needed provisions, with Snowflake and St. Johns experiencing particularly desperate conditions that year.

With church assistance, a grading contract was obtained for 5 miles of track along the continental divide approximately 150 miles east of the Little Colorado settlements, and on July 5 1880 about 40 men and 20 teams left Snowflake for the construction site (Fish n.d.a.:9; Peterson 1973:129).[3] On the basis of their grading contract, the settlers were able to secure supplies from merchants in Albuquerque, a portion of which was "sent back to the destitute settlements. By this means the colonies were relieved, and all felt very grateful after passing through one of the closest times for breadstuff that they had experienced in their settling of this country" (Fish n.d.a.:9).[4]

While work proceeded on the initial railroad contract, John W. Young, a son of the then late Brigham Young and director of the Mormon construction effort, obtained further contracts from the railroad company to grade an additional 100 miles of track through the Little Colorado region itself by July of 1881 (see McClintock 1921:192; Peterson 1973:130). In order to meet the time constraints placed on Young, the labor of over 500 men was needed to fulfill these contracts (Peterson 1973:131). A work force of this magnitude was beyond the means of the Little Colorado settlements, and additional labor had to be recruited from the numerous Mormon towns in southern Utah. Financially mismanaged, the subsequent construction venture became an economic failure in which most of the workers went unpaid.[5]

Although work on the first grading contract provided a strategic subsidy, forestalling widespread emigration, this subsidy was made available through a loss of labor which might otherwise have been invested in the forthcoming harvest.

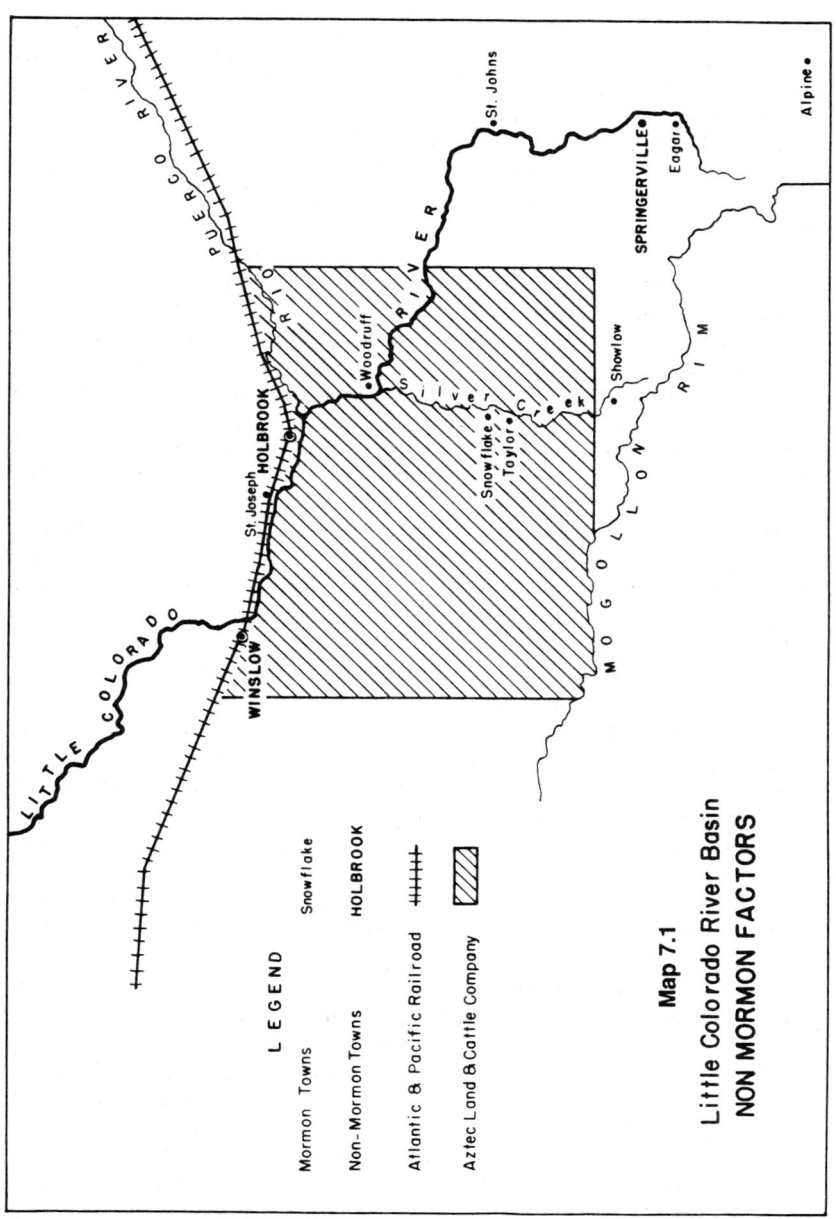

Map 7.1
Little Colorado River Basin
NON MORMON FACTORS

During the summer of 1880, settlements throughout the Little Colorado River Basin were left half empty (Peterson 1973:129), and the burden was left for "the few men remaining, together with the women and children,...to labor hard" in order to produce what little crops they could (McClintock 1921:192). While most workers on the initial railroad contract were paid (Peterson 1973:134), the lack of financial compensation for the labor removed from agricultural activities made the subsequent construction project a serious drain on the limited resources of the Little Colorado settlements.[6]

The coming of the railroad brought several benefits to the Little Colorado Mormon settlements in addition to the critical subsidy provided by the first grading contract. Supplies that previously required extended trips to either Kanab, Utah or Albuquerque (and later Socorro), New Mexico[7] (see Porter n.d.a.:60; Greenwood 1960:13; Tanner and Richards 1977:84) could now be more easily and more cheaply obtained at the newly-established town of Holbrook.[8] At the same time, the growth of two towns along the railroad line combined with the regional population increase and economic development stimulated by the railroad's arrival[9] generated a greater local demand for Mormon agricultural produce and made "peddling" a regular feature of the indigenous Mormon economy (Tanner and Richards 1977:79; Peterson, personal communication). The railroad also brought eastern commercial markets within reach of local producers.[10] As early as May of 1882, a shipment of 8,943 pounds of wool was sent from St. Joseph to Philadelphia, earning $1,4190 for that settlement (Warner 1968:22; Tanner and Richards 1977:84).

The economic boom which followed the railroad's arrival also increased the demand for local labor, a new opportunity which was readily exploited by Mormon settlers in the basin. Tanner and Richards (1977:79) list some of the non-farming occupations at St. Joseph subsequent to the arrival of the railroad. These included: "tending post office, stabling and feeding mail horses, shoeing horses for travelers, shearing sheep, repairing wagons, working on the railroad, cutting and hauling ties, building fence, hauling freight, and cutting and stacking hay."

Holbrook quickly emerged as the shipping point for all of northeastern Arizona,[11] with livestock and livestock products being the principal freight dispatched from this station (see Table 7.1). As early as 1881, the year of its founding, 300,000 pounds of wool were shipped from Holbrook. This was apparently the largest single wool shipment ever made by an Arizona firm (Wayte 1960:107). Following the arrival of the Aztec Company and the influx of more than 40,000 head of cattle (see below), between 10,000 and 20,000 head of cattle were shipped annually from Holbrook. Although complete figures are not available, Table 7.1 indicates that a steady increase occurred in the amount of produce (particularly livestock) exported from the region via Holbrook. By 1896, wool shipped through Holbrook had increased to 1,120,000 pounds (560 tons) (*Holbrook Argus*:6/19/1898). During this same year, nearly 20,000 head of cattle and over 22,000 head of sheep were shipped from Holbrook as well (*ibid.*).[12]

Table 7.1

Character and Amount of Freight
Shipped from Holbrook and Winslow, Arizona
on the Atlantic and Pacific Railroad Line
during 1885, 1888 and 1889
(in tons)

Item Shipped	Holbrook			Winslow		
	1885	1888	1889	1885	1888	1889
Grain	5	2	6	--	2	1
Hay and straw	1	--	6	--	11	--
Fruits and vegetables	2	--	--	--	--	--
Salt	19	13	3	1	--	1
Petrified wood	--	15	--	--	--	--
Stone, cement	--	1	--	--	--	--
Flour and millstuffs	4	4	--	3	--	--
Meats	--	6	--	--	--	--
Wines, liquors	1	--	--	1	--	--
Machinery, castings	--	4	--	--	--	--
Vehicles, tools	--	2	--	1	--	1
Livestock	402	1,161	2,248	43	970	345
Lumber, wood products	16	--	2	--	3	--
Furniture	7	16	19	5	26	1
Wool[a]	180	194	308	74	82	123
Hides, pelts and leather	42	19	25	8	7	9
Hardware	3	1	--	--	2	--
General merchandise	60	181	28	75	29	58
Oils	---	11	5	--	--	--
Railroad Co. materials	666	785	795	24,195	11,361	11,793
Total[b]	742	1,630	2,650	211	1,132	566

SOURCE: Guild (1891:62-65).
[a] Twenty-two (22) tons of wool were also shipped from St. Joseph in 1885.
[b] Totals exclude railroad company materials.

Table 7.2

Character and Amount of Freight
Delivered to Holbrook and Winslow, Arizona
on the Atlantic and Pacific Railroad Line
during 1885, 1888 and 1889
(in tons)

Item Delivered	Holbrook			Winslow		
	1885	1888	1889	1885	1888	1889
Grain	184	205	675	51	126	111
Hay and straw	--	68	77	--	88	71
Fruits and vegetables	30	61	96	32	41	36
Bar and Sheet Metal	29	--	15	--	1	--
Coal	139	126	139	143	206	341
Salt	14	38	27	32	44	27
Brick, stone and cement	2	1	3	8	15	1
Flour and millstuffs	337	393	407	50	70	63
Meats	1	34	44	4	7	2
Wines, liquors	67	53	61	30	54	53
Machinery, castings	14	6	42	1	2	--
Agricultural implements	5	12	4	--	--	--
Vehicles, tools	9	42	36	2	3	2
Livestock	3,903	628	80	20	11	15
Lumber	328	136	60	147	209	127
Furniture, household goods	12	51	33	19	25	15
Oils	11	20	38	3	20	18
Water, ice	--	--	15	3	24	75
Wool	--	--	--	--	--	--
Hides, pelts and leather	1	--	--	--	--	--
Hardware	58	55	53	6	10	6
General merchandise	895	540	646	200	486	323
Ores	--	--	--	--	--	--
Railroad company materials	1,366	964	342	70,606	20,098	36,357
Total[a]	6,039	2,469	2,551	752	1,442	1,286

SOURCE: Guild (1891:66-71).
[a] Totals exclude railroad company materials.

External Impacts   171

Regional population growth is reflected in the increasing importation of such basic commodities as grain, flour and general merchandize throughout the 1880's (see Table 7.2).[13]

While regional economic development brought benefits to local Mormon settlements, the advent of the railroad and the growth of such non-Mormon towns as Holbrook and Winslow also brought competition to local Mormon producers. Due to the reduced costs provided by railroad transportation, wheat--always difficult to grow in the lower valley--could be obtained at St. Joseph as cheaply (and more reliably) by purchasing it from outside sources as by raising it indigenously (see Tanner and Richards 1977:77-78). Moreover, the imported wheat was considered by many to have been of better quality than that raised locally (*ibid.*). Consequently, many settlers at St. Joseph began to buy imported flour, and the *Holbrook Argus* (4/23/1896) reported that "farmers along the Little Colorado River are converting their ranches into alfalfa fields as rapidly as possible, the acreage being almost doubled this year."[14] The increasing abandonment of wheat farming along the lower valley forced the grist mill at St. Joseph to close,[15] and those who continued to grow their own grain had to travel to Shumway to have it milled.[16]

Although the business competition generated by the railroad's arrival undermined local self-sufficiency--particularly at St. Joseph which not only existed in immediate proximity to the competition, but which was also located within a marginal habitat for agriculture--this same competition underlay the origin of the regional Mormon resource-flow system based on tithing redistribution. The establishment of a board of trade, the formation of the ACMI and the more effective redistribution of tithing resources that these institutions provided was largely a product of the economic competition generated by the arrival of the railroad.

Two additional developments resulted from the railroad's penetration of the Little Colorado River Basin that had important implications for the settlement process. The first involved the arrival of the Aztec Land and Cattle Company. The second concerned the intense political conflict with non-Mormon interests in the basin. These two developments jeopardized the survival of several settlements and, thus, posed serious threats to the success of the colonization effort. They were eventually overcome because considerable resources were provided to local settlements though the church organization.

The Aztec Land and Cattle Company

Under financial pressure stemming from the unprofitability of its newly-constructed western line, The Atlantic and Pacific Railroad Company sought buyers for 5,424,800 acres of land granted to it in conjunction with the railroad contract of 1866 (see Kennedy 1968:2-3). One million of these acres were purchased at $.50 per acre by the Aztec Land and Cattle Company,[17] a consortium founded in New York City in 1884 by eastern businessmen and Texas ranching interests.[18] The land claimed by the Aztec Company included every other section extending in longitude from 12 miles east to 50 miles west of Snowflake for a depth of 50 miles south of the railroad line.[19] By owning every other section and by maintaining control of all critical water sources throughout its domain, the Aztec Company effectively monopolized over 2,000,000 acres of range land. They, therefore, removed a substantial resource supply from local utilization.

The Aztec Company was a direct product of the deteriorating range conditions in western Texas during the mid-1880's.[20] Open range, rising meat prices and mild weather created a highly speculative investment atmosphere during the early 1880's that attracted considerable capital (much of it from English and Scottish cattle syndicates) and led to the importation of millions of head of cattle into the arid West Texas region. By 1885, however, the bubble had burst; the introduction of barbed wire and windmills, together with the passage of legislation favorable to small farmers, spelled the end of the open range. In order to recoup their investments before the complete encroachment of farmers and the subsequent enclosure of grasslands occurred, ranchers in West Texas dangerously overstocked their ranges.[21] "Overstocking of the range had so reduced the grass that either a drought or a hard winter would bring disaster" (Webb 1931:237).

Disaster did strike; by 1885, thousands of cattle had died and thousands more existed half-starved on a barren range.[22] A drought, which began in 1883, led to a fall of prices in 1884 that was followed by a crash in the beef market in 1885. Unable to sell their livestock at profitable prices, ranchers searched for a new range where cattle could be maintained until the market rebounded.[23] The lush grasslands of northern Arizona, combined with the financial difficulties of the Atlantic and Pacific Railroad, made the formation of the Aztec Land and Cattle Company and the relocation of thousands of head of Texas cattle on this cheaply-acquired land an attractive opportunity to many ranchers in western Texas.[24]

Although it experienced little difficulty finding livestock to purchase, the Aztec Company encountered considerable problems transferring those animals to their newly acquired range in northern Arizona.[25] In spite of these difficulties, the company imported between 33,000 and 40,000 head of cattle into northern Arizona by the close of 1887, which quickly grew to a herd of about 60,000.[26] To manage their investment, the Aztec Company also imported some 100 (mostly Texas) cowboys, supplies and over 2,500 horses. The Hashknife Outfit,[27] as these cowboys came to be known, soon earned a notorious local reputation (see below).

Despite one report to the stockholders in 1877 recommending a "liberal policy towards persons taking a homestead" on Aztec Company land (so as to counter the negative local attitude that had developed towards the company) (see *Hoof and Horn*:2/3/1887), an exclusionist policy prevailed.[28] Occupying every possible waterhole and stationing men along the borders of its territory, the Aztec Company effectively excluded most competitors. Dealing sternly with trespassers,[29] this exclusionist policy frequently resulted in violent confrontations between cowboys employed under the Hashknife brand and individuals who encroached upon company land. In addition, the Little Colorado River Basin provided yet

External Impacts 173

another setting for the classic struggle between cattle and sheep interests in the American West.[30] In 1887, Hashknife cowboys became involved in a struggle between the Grahams and the Tewksberrys, cattlemen and sheepmen respectively in the Tonto Basin to the southwest. The conflict between these two families quickly erupted into the Pleasant Valley War which left 21 men dead.[31]

The destruction wrought upon the overcrowded ranges of the Little Colorado River Basin during the 1890's has already been discussed (see Chapter 4). As previously happened in West Texas, the devastation came quickly and resulted from the speculative nature of the nineteenth century cattle industry. Successive droughts, repeated economic crises, and declining cattle prices[32] during the 1890's produced dangerously overstocked ranges. Maintaining more than 60,000 head of cattle on the range while waiting for better prices, the Hashknife outfit reportedly branded as many as 52,000 head in one year (Kennedy 1968:20). When repeated years of drought were followed by the severe winter of 1898, thousands of cattle perished. The extent of the loss is clearly illustrated by the following account which appeared in the *Holbrook Argus* (12/16/1899):[33]

> We had heavy losses in our section this last year from drought. Along the Santa Fe Pacific the loss has been fully 50 percent. The Aztec Company expected to gather between 35,000 and 40,000 head and they cleared up about 16,000 head. The big outfits are all pulling out of our part of the territory and they are going for good.

As Kennedy (1968:21) clearly notes, "the Texans fled to Arizona to escape the effects of their malpractices. They proceeded to repeat the process in Arizona. Drouth and range deterioration followed as a matter of course." The effect of the arrival of the Aztec Company on the Little Colorado Mormon settlements were immediate and severe. Excluded from the ranges they formerly exploited and experiencing increased competition on remaining lands, Mormon settlers, many of whom had raised either cattle or sheep, suffered economically. Shortly after the Hashknife outfit established its territorial claim, the Eastern Arizona Stake herd was described as being in a "sickly condition...hardly able to maintain an existence" (Fish n.d.a.:19). Fish (*ibid.*:21) further noted that as early as 1877, "many of our people were disposing of their stock as fast as possible".

Equally devastating from the perspective of the Little Colorado Mormon settlers was the rampant lawlessness which followed the Aztec Company's arrival. Formerly isolated, subject exclusively to the repercussions of their own actions, the various Mormon settlements soon found themselves in the midst of a bawdy cattle frontier with all the notorious events and characters that have made Dodge City and Abilene famous names in Western lore.[34] One report claimed that "nearly all of the valuable horses in the Snowflake part of the country were driven out, the settlers thus losing thousands of dollars" (see Jensen n.d.b.:4/2/1887).[35] Although the depredations of the Hashknife cowboys were felt throughout the basin by Mormons and non-Mormons alike, the isolation of their towns and the aloofness which they maintained made the Mormons a favorite target of the rustling and violence perpetrated by this outfit.

While the drains imposed by range competition, cattle rustling and violence bore heavily on Mormon settlers in the basin, the most acute threat presented by the arrival of the Aztec Company concerned the possible loss of the very lands upon which these setters had invested so much of their resources during the preceding ten years. Since the railroad grant acquired by the Aztec Company pre-dated all Mormon claims in the basin, any town located on Aztec Company land had to negotiate with the company for the purchase of that property. Three Mormon settlements--Snowflake, Taylor and Woodruff--were immediately threatened with the loss of their land.[36]

After failing to persuade the government to grant the railroad other lands in lieu of those already occupied by Mormon settlers in the basin, Jesse N. Smith, the Eastern Arizona Stake President, went to New York to negotiate with the Aztec Company for the purchase of 7 sections of land near Snowflake and Taylor. A deal was consummated in which Smith agreed to a price of $4.50 per acre. One-fifth was paid down, with the balance to be paid in 4 yearly payments at an interest of 6% per annum, payable every 6 months (Fish n.d.a.:66).[37] The settlers at Woodruff, who were resident on land still owned by the Atlantic and Pacific Railroad, were forced to pay $8.00 per acre for the one section of land which they had to purchase (*ibid.*).[38] In both cases, church headquarters in Salt lake City assisted in the negotiations and forwarded the money necessary to cover the downpayments.

A meeting of the farmers from Snowflake and Taylor was held in which it was decided that those who were fortunate to own government land would share some of the cost imposed upon those who by chance settled on railroad property, since all land owners in these two settlements were interdependent in their mutual reliance upon a common irrigation system. The former group agreed to pay $6.25 per acre towards the purchase of the 7 sections acquired, while the latter paid $12.50 per acre.[39] Although distributing the costs where possible certainly eased the burden carried by individual settlers, the drain was by no means eliminated, as Fish (n.d.a.:66-67) noted.

> This we considered was very dear, but it was the best we could do unless we abandoned our improvements. The company had taken advantage of this to charge us such high prices for it. Our taxes are already very high[40] and this seemed a terrible sum for a few families who were already struggling with poverty to raise.

In order to assure that the settlers would gain clear title to their lands quickly, the church paid both the Atlantic and Pacific Railroad and the Aztec Company the full purchase price for all the land acquired at Snowflake, Taylor and Woodruff in 1891.[41] Although settlers in these towns were expected to repay the church the money it had forwarded, the drought and the consequent depression which prevailed in the basin during the 1890's made repayment an economic impossibility for most. Accepting the inevitable, the church applied these debts elsewhere and, as Peterson (1973:175) appropriately remarks, "the entire balance ...due the church thus served as a subsidy to the continued development of the colony".[42]

Drought, range deterioration and falling cattle prices during the 1890's, combined with heavy losses of cattle from both starvation and rustling by its own cowboys, forced the Aztec Company into bankruptcy in 1900. After only 16 years of operation the company was obliged to liquidate its extensive holdings, thus ending the era of speculative cattle ranching in the Little Colorado River Basin.[43] During its brief reign, however, the Aztec Company had a devastating impact on ranges within the basin and, thus, on those populations that depended upon these ranges for their existence.

While livestock absolutely dominated the regional economy following the Aztec Company's arrival, the advent of the railroad stimulated a variety of economic developments in the basin in addition to livestock production. Concurrent with the region's economic development came increasing competition over access to and control of resources which facilitated the advancement of specific business interests. Most notable was the competition that emerged over control of the governmental apparatus established in 1880 to administer the newly formed Apache County.[44] Throughout most of the 1880's the prevailing political climate in Apache County was not unlike that which has existed in other frontier regions where vested business interests have attempted to monopolize political office.

Apache County Politics

Pursuant to their goal of remaining self-sufficient and aloof from "Babylon", (cf. Little Colorado Stake:8/2/1879), the earliest Mormon pioneers in the Little Colorado River Basin eschewed active involvement in the political affairs of the region. However, as in previous circumstances surrounding Mormon settlement (cf. O'Dea 1957:41-75), the size and rate of the Mormon influx posed a direct political and economic threat to other local interests. Consequently, despite their initial apoliticism, Mormon settlers along the Little Colorado quickly became the object of pointed and often violent persecution.[45] The intent of this harassment was advanced by, but did not originate from, the larger national effort to disassemble the Mormon Church and rid the country of polygamy and an independent theocratic state.

Mormon political problems along the Little Colorado emerged within five years of the founding of their initial settlements and developed largely in conjunction with their colonization of the upper reaches of the Little Colorado River where two non-Mormon towns--St. Johns and Springerville--had already been established. Conflicts arose initially between Mormon settlers and the indigenous Hispanic population at St. Johns. The latter were directly and immediately threatened by the advent of Mormon settlement there and by the implications that increasing Mormon immigration posed for competition over farmland and irrigation water. Adverse relations quickly expanded to include Mormon interactions with Anglo business interests centered at the same town. The near-frantic influx of Mormon settlers into St. Johns during the winter of 1879-1880 and again in 1881 and 1884 likely appeared as nothing less than an invasion to those who feared being overwhelmed in the rush.

To Hispanics at St. Johns, most of whom subsisted primarily by farming and sheepherding, the successive waves of Mormon settlers portended a loss of control over water and land resources that they had long assumed were theirs. The concerns of St. John's Hispanic population did not temper Mormon immigration; however, some settlers were at least partially cognizant of the roots of their conflict with this group.

> The Mexican people saw us surveying the land adjacent to their town on the west; they saw new settlers coming to swell our ranks. I doubt that they realized we had bought this land with the view of making homes there. I am sure that they did not realize that we had no intentions of molesting them; rather they looked upon us as enemies, who had come to encroach upon their old "San Juan", settled by them in 1873. The Mexicans resented us and we did not blame them very much. Their "squatters rights" had not been properly respected by those who sold the land to our people. (Udall n.d.:77)

The construction of storage reservoirs by the Mormon settlers at St. Johns during the 1880's soon alleviated the seasonal water shortage at this location and removed a major source of conflict with the Hispanic population there.[46] Although the conflict between Mormons and Hispanics subsided with time, the animosity which surfaced between Mormons and certain non-Mormon business interests in the region intensified (see Udall n.d.:114-115). The St. Johns Ring, as this political-mercantile collusion came to be known, was determined to acquire and maintain control of the newly established governmental apparatus in the basin.[47] The potential voting block represented by the rapidly increasing Mormon population posed a threat to that ambition.[48]

Beginning with the initial county elections held in 1879, the Ring practiced covert and overt electoral fraud with increasing impunity as it consolidated its power throughout the region. Ballot stuffing, rigged vote counts, control over election officials, barring Mormons from polling places, and refusing to seat Mormon candidates when elected were some of the techniques employed by the Ring to effectively disenfranchise and disempower the growing regional Mormon population.[49] Fish's (n.d.a.:31; see also Peterson 1973:226) account of his attempt to vote in the 1882 election illustrates how quickly the Ring was able to consolidate its power.

> On November 7th I went down to Holbrook to attend the election there and watch and see if I could prevent any frauds that were being perpe-

trated at the elections by the ring. I soon found that I could do but little. A man stood in front of the polls armed as if for war and I was not allowed to come nearer than fifty feet of the polls and was threatened with arrest even before I had spoken a word to anyone. I went up to vote and was marched back by the constable John Conley who again threatened me with arrest. In the meantime others who belonged to the party[50] in power went up and voted without being molested.

The Ring also barred Mormons from jury duty and threatened their schools by making it impossible for Mormons to pass teacher-qualifying examinations. Furthermore, by controlling all judicial appointments the Ring effectively obstructed Mormon efforts to redress their grievances through legal channels (cf. Fish n.d.a.:52-53; Peterson 1973:226-227). The overt persecution of Mormons by the Ring encouraged others to exploit the situation, and a virtual license existed for those who wanted to advance their own interests at Mormon expense. Claim jumping (cf. Fish n.d.:52; Peterson 1973:170), particularly at St. Johns[51] and in Round Valley[52] (cf. Udall n.d.:92; Greenwood 1960:97-98), theft of livestock[53], and even violence[54] (cf. Fish n.d.a.:52-53; Udall n.d.:92; McClintock 1921:190; Peterson 1973:170) became more than uncommon occurrences visited upon Mormon settlers in the basin.

Although hostility towards Mormons occurred throughout the basin, the heaviest burden was borne by those Saints residing in or near St. Johns. St. Johns was the county seat, the center of operations for the St. Johns Ring, and the area with the heaviest concentration of Hispanic settlers.[55] Although some of the problems he mentions were experienced by Mormon settlers throughout the basin, Fish (n.d.a.:52-53) recorded several comments regarding the situation at St. Johns that are usefully quoted at length.

> The people of St. Johns have met with very bitter opposition from our enemies, both whites and Mexicans from their first settling there until within the last few months.
>
> The saints have been deprived of their rights both politically and legally by those in power in the county and territory. The people have been robbed of their water privileges, their land has been jumped, their stock stolen, leaving many of the brethren without teams.
>
> They have been insulted and abused until at times it seemed unbearable.
>
> They have been forced contrary to law to assist their neighbors in building an expensive school house. The taxes levied upon the people for this purpose amounted to more than $900.00.
>
> The people have been deprived of the privilege of voting merely by the diction (sic.) of the local judge of election. Their school district has been taken from them without due process of law; and the people have been

forced to accept such teachers as the ring sent and to associate their children with those of low filthy habits in the school room, or employ and pay their own teachers and be deprived of the school funds that were justly and legally due them; the people chose the latter.[56]

They have fenced and built in our streets, and when we have applied to the courts for redress of these grievances, they have invariably decided against us.

There has been quite a number of men called to St. Johns, of whom perhaps two thirds have come, numbering about 250. Of that number there are about 140 remaining some being discouraged with remaining because of the country and some with the people, left the place and have gone to other places.

As a result of continuing political harassment, the circumstances of Mormons at St. Johns soon became desperate, and outside help was deemed necessary to hold the colony.[57]

At the time of the lot jumping, St. Johns received aid and comfort from the various wards in the stake agreeable to the instruction of the president of the stake, by many of the brethren going to the aid of the brethren and helping them to hold their lands and locations, some of the brethren remained at St. Johns for several months. About this time the feeling was so bitter against the saints that the saints felt under the necessity of carrying firearms. (Fish n.d.a.:53)

As had frequently happened in other situations considered worthy of special support, assistance from other settlements in the region was supplemented by resources obtained from Mormon sources outside the basin. As Fish (*ibid.*) reports:

There has been considerable assistance rendered the St. Johns ward by the church and by the people of the stakes of Utah. In the spring of 1885 President John Taylor issued a Tithing Office Order for $1,000.00 and there was collected from the different stakes in Utah $1,187.00. Said amounts were appropriated and used by the settlers of St. Johns for breadstuff and seed grain.

Outlawed by the Anti-Bigamy Act of 1862 and the even more threatening Edmunds Act of 1882, polygamy assumed increasing prominence as an issue in the conflict between Mormons and non-Mormons in Apache County,[58] culminating in the judicial raid of 1885. As a result of these polygamy raids, several prominent Mormons were arrested, convicted of bigamy, and incarcerated in federal penitentiaries (see Peterson 1973:222-223; Fish n.d.a.:18-19). Many others fled to Mexico early that year in order to escape prosecution.

The repercussions of the judicial raid extended beyond the suffering of specific individuals. Because those most adversely affected by anti-polygamy legislation were church leaders and others heavily involved in the complex Mormon regional administrative organization, the judicial raid directly affected the integrated system of resource flows through which the various settlements mitigated the potentially destabilizing consequences of environmental variability (see Chapter 6). The removal of church leaders resulted in the loss of extensive administrative experience. In addition, since those who left requested and received sufficient financial resources to sustain them while residing outside the territory, their departure imposed a serious drain on local settlements and on the regional system of resource redistribution, especially the ACMI. Fish (n.d.a.:18-19) reports that by the close of 1885 most co-operative activities throughout the Eastern Arizona Stake were in desperate condition. Even the Snowflake co-op store, which was in better condition than most, was doing a business "not quite so extensive as it had been before the judicial raid." Fish (*ibid.*) added:

> No institution felt the effect of this move more than the Arizona Cooperative Mercantile Institution. The President, Vice President, superintendent, secretary and the majority of the board of directors were among the number who left the territory. This with the heavy draw by the shareholders who left, and the stagnation of business was a very heavy blow, and the institution came very near failing, but by careful management it was again put on a sound footing.[59]

The polygamy issue was not the cause of local hostilities directed at Mormon settlers. The source of this animosity was indigenous and resulted from the rapid immigration of Mormons into the basin and the competitive threat that this portended to growing commercial interests in the region. The polygamy issue merely represented a "bridge-action" (see Bailey, 1957) by which these local individuals exploited outside resources to advance their own interests.

The political climate for Mormons in the Little Colorado River Basin began to improve after 1887. During that year, a Grand Jury met at St. Johns which handed down several indictments against individuals associated with the St. Johns Ring (Fish n.d.a.:22). Although the work of this first body was cut short, a second Grand Jury convened in 1892 which indicted 22 persons. This resulted in four convictions, including that of Sol Barth the leader of the Ring. With Barth and others in jail, the Ring crumbled. However, Mormon political prospects had already begun to improve before the indictments of 1892. During the 1888 election in which the People's Party offered a strictly anti-Mormon slate of candidates, a full Democratic ticket including two Mormons was elected (Fish n.d.a.:66). Three Mormons were subsequently elected to office again in 1890: one to the Territorial Legislature, one as County Recorder, and the other as County

Treasurer (Fish n.d.a.:68). These political gains within the county occurred simultaneously with supporting changes at the territorial level (see Peterson 1973:234-240). Regarding the 1890 election, Fish (n.d.a.:68) reported that "this election created less feeling against the saints than usual and things were left quiet for our people." For the closing decade of the century, the political stress imposed on Mormon settlers had largely disappeared.

Church Resources

The Mormon Church was perhaps the most prominent of the competing forces influencing Mormon settlement of the Little Colorado River Basin. Moreover, while other external influences potentially threatened the settlement process, church resources were invariably used to advance the colonization effort. The church, with its substantial assets, frequently became the source of material resources which determined the difference between the continuation or abandonment of a specific settlement.

A complete list of the subsidies provided to the Little Colorado settlements through church headquarters would require an unnecessary recapitulation of data already presented in those chapters specifically examining the settlement process. The very settlement of the Little Colorado River Basin stemmed from a decision made in Salt Lake City, and the necessary preparations were carried out under church supervision and at church expense. In addition, beginning with its donation of both a grist and a saw mill during the initial year of colonization, the church repeatedly provided critical subsidies that enabled individual settlements to construct and reconstruct irrigation systems and other public structures, to weather poor or nonexistent harvests, to purchase and repurchase lands from individuals or corporations with prior claims, and to endure the financial losses attending political harassment.

The church also furnished non-monetary subsidies to its satellite towns within the basin in the form of legal services, various skills and experiences, and in conducting negotiations for railroad grading contracts and for the purchase of land. It also provided local settlements with a hierarchical organization which effectively integrated them into a regional resource-flow system linked through parent institutions to Mormon towns in Utah and elsewhere. The Little Colorado settlements were also frequently visited by Apostles and other church authorities who offered valuable advice and who remained continuously appraised of developments within the basin. In addition, through its institution of the "mission," the church provided these settlements with a more or less steady supply of manpower which, owing to heavy emigration, frequently threatened to fall short of community maintenance requirements at specific locations.

The overriding importance of the church connection lay in the fact that available church resources were invariably applied to advance the settlement process. The hierarchical and integrated structure of the larger church organization, with its clear channels of authority and its interregional connections, facilitated a rapid and flexible application of substantial resources to a variety of local

External Impacts   181

problems of varying spatial dimensions. The saving and centralization of surplus productivity in the form of tithing stocks assured that a ready supply of local resources was generally available which could be immediately directed into those specific channels deemed critical to the colonization effort.

While it is true that in its effort to limit Gentile settlement within its domain the church may have condemned many of its faithful to a rather bleak and strenuous existence in the Little Colorado River Basin,[60] from the perspective of examining the evolution of agricultural communities in this region it is clear that the church also provided critical subsidies without which several individual settlements--and likely the entire colonization process--would have failed.[61] Even with continued church support, Obed, Old Taylor, Brigham City and Sunset succumbed to the ravages of the Little Colorado River, while other settlements had to be abandoned a higher elevations. Without church support, it is highly unlikely that either St. Joseph or Woodruff would have survived, and continuous calls for new settlers were needed to bolster the declining population at St. Johns as a result of the political climate which prevailed there. Even Snowflake and Taylor benefitted from the church's assistance in obtaining the first railroad grading contract, in the purchase of their land from James Stinson, and in the re-purchase of this same land from the Aztec Company.

CONCLUSION

Recurring cycles of railroad construction, land speculation, immigration, population growth and economic expansion followed by years of resource overexploitation, environmental deterioration, economic recession, emigration and population decline characterized the westward expansion of the American frontier (cf. Webb 1931; Shannon 1945; Gates 1960). In this regard, the Little Colorado River Basin proved no exception. Mormon communities there sustained serious challenges to their survival from the increased instability that the arrival of this frontier engendered.

Practically every Little Colorado Mormon settlement was affected by the increased competition which followed the railroad's arrival. The settlers at Snowflake, Taylor and Woodruff came perilously close to losing their land, while the residents at St. Johns and in Round Valley suffered heavy financial drains due to intense conflict with local non-Mormon interests. In addition, Mormon settlements throughout the basin endured losses in productivity due to the expropriation of rangeland by the Aztec company, the range deterioration produced by overstocked ranges, the lawlessness perpetrated by the Hashknife cowboys, and the intense anti-Mormon political climate which prevailed during the 1880's.

Conditions in the basin following the arrival of the railroad caused population size to decline in nearly every Mormon settlement after 1885 (see Table 2.6).[62] In most cases, populations did not regain their former numbers until after the drought of the early 1890's. The exclusionist policy of the Aztec Company and the subsequent overcrowding and deterioration of rangeland produced a significant decline in livestock productivity for most Mormon settlements in the basin (see Table 2.5).[63] Faced with the competition presented by their immediate proximity to both the railroad line and to the town of Holbrook, settlers at St. Joseph even abandoned farming in favor of alternate means of subsistence.[64] By 1900, not one person from this town declared farming --the mainstay of the ideal, self-sufficient nineteenth century Mormon community-- as their principal occupation. Insurmountable competition had caused the abandonment of (i.e., exclusion from) previous exploitative activities.

Although the various instabilities that accompanied the expansion of the American frontier could not be avoided when this frontier penetrated the Little Colorado River Basin, the stresses imposed on local Mormon settlements were substantially mitigated by their church affiliation. The ability of several specific towns to successfully weather the economic and political uncertainties of the 1880's is directly attributable to material subsidies provided through the hierarchical church organization which extended into the region. As with other territories colonized by Mormon pioneers,[65] the stresses endured by Mormon settlers along the Little Colorado were considerably less than they might have been had the resources of the larger church organization not been made available to subsidize the colonization effort.

Frontier communities are notoriously unstable.[66] Rapid immigration and population growth, the introduction of populations and exploitative techniques largely unadapted to novel environmental conditions, and the intense competition which often accompanies the scramble for control over newly-available resources typically produces community instability in recently settled territories. However, such instability generally declines as increased productivity, evolving adaptations, decreased competition, and the evolution of regulative functions enhance local stability.

In the evolution of human communities, factors traditionally classified as economic, political or social may produce developmental effects comparable to those defined as ecological in nonhuman communities. Instability or reduced productivity may have the same impact on community development regardless of whether such conditions derive from unpredictable features of the natural environment or from the uncertainties attending economic or political competition. The following chapter will examine the principal features associated with community development during Mormon colonization of the Little Colorado River Basin in terms of the ecological model presented in Chapter 3. The goal will be to provide a consistent, synthetic and parsimonious explanation for various features of the settlement process discussed in the previous chapters, including the impact of those external factors that would not have been traditionally included within an ecological analysis.

NOTES

1. "Costs" here include monetary payments, human labor expended to establish and maintain newly acquired townsites and the drain imposed on the resources of donor settlements charged with the duty of subsidizing the colonization of new locations. The urgency with which colonization of the basin was viewed--at least by church leaders-- is illustrated by the following comment attributed to Wilford Woodruff, a Mormon Apostle during the settlement of the Little Colorado River Basin (Udall n.d.:74):

> We must hold St Johns at all costs, or it will become a second Carthage to our people in northern Arizona.

Carthage, Illinois was the town at which the Mormon prophet Joseph Smith was assassinated by an anti-Mormon mob. The events surrounding Smith's death generated the westward migration of Mormons and the founding of Salt Lake City as the new center of Zion (see O'Dea 1957:53-75; Arrington 1958:17-18; Leone 1979). See also footnote 60 below.

2. Freighting to Fort Apache was vigorously pursued by the ACMI, which after its move to Holbrook in 1888 became one of the principal suppliers to this military post. In 1883, over 880,000 pounds of supplies were shipped by wagon from Holbrook to Fort Apache (*Apache County Critic*:6/24/1884), and this figure grew dramatically to about 3 million pounds annually during the 1890's (see Jensen n.d.c.: 2/27/1894).

3. Smaller contingents of men and teams were also dispatched from other settlements in the basin (see Peterson 1973:129).

4. The Minutes of the Eastern Arizona Stake (page 109; see also Peterson 1973:126) indicate that in June of 1880 a deficiency of about 10,000 pounds of grain existed below that needed to maintain the stake population until the following harvest. The shortage of grain was exacerbated by the inadequacy of the 1880 harvest, which resulted largely from the heavy drain of labor involved in the railroad construction (see Fish n.d.a.:9).

5. See Peterson (1973: 129-136) for a discussion of the circumstances surrounding the economically unsuccessful railroad contracts.

6. About $2,500 in surplus goods from Young's misadventure did eventually become recycled to form part of the capital that inaugurated the formation of the ACMI in 1881 (see Fish n.d.a:10; Peterson 1973: 137).

7. Greenwood (1960:13) states that a roundtrip journey by wagon from the Little Colorado settlements to Socorro, New Mexico (the same distance as to Albuquerque) required 15 to 20 days. Tanner and Richards (1977:84) claim that return trips to home towns in Utah were reduced for many St. Joseph residents from 6 weeks to 2 or 3 days by the arrival of the railroad.

8. The importance of the railroad connection to the settlers at St. Joseph is clearly illustrated by the scenario of events that transpired within 8 months of the railroad's arrival in the basin. So desperate was the situation at St. Joseph in 1881 that a team manned by two men was dispatched from here in September of that year to obtain flour at Sanders, Arizona even though the railroad was to arrive at this lower valley settlement within two weeks (see Chapter 2). A large shipment of flour was received at St. Joseph again in November. Another carload of flour and grain was received there in January of

1882, and still more supplies were obtained from Holbrook in February, and then again in March, of the same year (Tanner and Richards 1977:84). This swift exploitation of the railroad as a source of basic supplies, particularly foodstuffs, underscores the fundamental dependence that St Joseph (and several other settlements) had upon externally-obtained resources and the immediate benefit wrought such towns by the railroad's arrival. A frequently required resource had simultaneously experienced a reduction in cost and an increase in availability. According to Tanner and Richards (1977:78), freight charges on produce shipped from Kanab, Utah could more than double the cost of items obtained from the railroad depot at Holbrook.

9. Established in 1881, the population of Holbrook increased to around 200 persons as early as 1884 (Wayte 1960:114) and to over 600 persons by 1900 (Bureau of the Census 1900). Founded in the same year as Holbrook, Winslow's population grew to 363 persons in 1890 and 1,305 persons in 1900 (*ibid.*). Overall, the size of the non-Indian population in the Little Colorado region increased from 3,464 persons in 1880 to 6,778 persons in 1900 (*ibid*).

10. Conversely, this newly-created access to eastern markets also opened the Little Colorado River Basin to non-local ranching interests, culminating in the arrival of the Aztec Land and Cattle Company (see below).

11. Although Winslow was the larger of the two towns along the railroad line, Holbrook served as the regional shipping and marketing center. From its beginning, Winslow has functioned as a terminal point for the railroad and, due to its peripheral location with regard to other towns in the basin, has performed a role in the regional hierarchy of towns incommensurate with its size. Holbrook's commercial pre-eminence within the basin had been maintained until very recently. With the growth of trucking as a primary technology of resource distribution and the rapid development of metropolitan centers such as Phoenix and Tucson in southern Arizona, alternate traffic patterns have emerged within the basin. As a result, Showlow, because of its southern location and greater proximity to the metropolitan centers of southern Arizona, has displaced Holbrook as the region's commercial center.

12. At an average weight of approximately 750 pounds, a shipment of 20,000 cattle would have totaled 7,500 tons. Twenty-two thousand sheep, averaging 33 pounds each, would have amounted to about 360 tons, producing a total livestock shipment out of Holbrook in 1896 of around 7,860 tons. (Compare with Table 7.1).

13. The totals for certain categories of items imported in 1885 are higher than in subsequent years. The reason for this discrepancy is that 1885 was the year in which the Aztec Company imported over 40,000 head of cattle, about 100 cowboys and the necessary supplies (including horses) to establish its extensive operation in the basin.

14. Alfalfa (luscerne) is used as a supplemental feed for livestock. The initial adoption of this cash crop among settlements in the lower valley suggests the difficulties attending the more diversified, traditional agriculture employed in this area and, therefore, of the greater viability of specialized cattle feed production for settlers in these towns. The reduced labor demands associated with alfalfa production also enabled participation in other economic activities.

15. The collapse of Sunset, Obed, Brigham City and Old Taylor meant that an insufficient Mormon population remained in the lower valley to sustain even a single grist mill there in the face of competition.

16. Shumway, which is located a few miles south of Taylor on Silver Creek, is twice the distance from St Joseph as was Sunset. Grist was, therefore, generally taken to Shumway to be milled during trips to Snowflake to attend the stake conferences.
17. The Aztec Company actually gained title to 1,059,560 acres of land, the additional acreage being included at no charge (Kennedy 1968:5).
18. Kennedy (1968:3) lists the following stockholders in the Aztec Company at the time of its formation:

The Atchinson Railroad	$215,000.00
Members of the Atchinson Board of Directors	200.00
J. W. Seligman and Company	214,000.00
Persons connected with Seligman firm	4,000.00
Others (mostly Texas ranchers)	502,000.00

The Aztec Land and Cattle Company was the third largest ranch in North America during the late nineteenth century (Dreyfuss 1972) and was the direct successor of the Continental Cattle Company established along the Pecos River in western Texas in the late 1870's. The Continental Cattle Company was the original owner of the hashknife brand made famous by the Aztec Company (Wayte 1960:134; see below).

Considerable interlocking business interests were represented in the formation of the Aztec Company. The J. W. Seligman Company was a banking firm and principal creditor to the Atlantic and Pacific Railroad that very likely hoped to recoup its loan to that firm through its investment in the Aztec Company. Edward Kinsley, also an original stockholder, served as a trustee for the railroad in disbursing its grant lands and similarly had outstanding debts owed him by the company. In addition, in 1897 a much expanded Atchinson, Topeka and Santa Fe Railroad later purchased the Atlantic and Pacific Railroad for $12,000,000, giving itself a through line to the Pacific Coast and thus becoming one of the most powerful railroad companies of the day (see *Winslow Mail*: 5/8/1897). The Building of the railroad and the introduction of the Aztec Company clearly brought the Little Colorado River Basin within the purview of the highly speculative range cattle business of the late nineteenth century.

19. The actual grant to the railroad was for a depth of 40 miles south of the track An additional depth of 10 miles was claimed by the railroad as lieu land to compensate for territory previously claimed elsewhere along its right of way. Although several years were to pass before the Aztec Company gained full legal title to all of the land that it claimed, the company immediately commenced its operations as if it did, in fact, have a clear title to this land (see Kennedy 1968:5).
20. See Webb (1931:233-260) for a discussion of the boom and bust nature of the cattle industry in western Texas during the 1880's. Between November of 1885 and February

of 1887, *Hoof and Horn*, a weekly publication of news relating to ranching concerns, contained numerous articles reporting the condition of ranges in western Texas as this bore on the importation of cattle into northern Arizona.

21. Webb (1931:237) reports that one ranch near Fort Worth, Texas stocked 25,000 cattle on a range of only 100,000 acres. Four acres was inadequate to support a steer in arid western Texas As Webb notes, "in the spring of 1883 the roundup brought in 10,000 head; 15,000 dead cattle on the range told the rest of the story".

22. One article in *Hoof and Horn* (5/20/1886) illustrates the extent of the devastation wrought upon both cattle and range in western Texas. Quoted in part from an article in the *Chicago Times*, *Hoof and Horn* presented the following report:

> The plains of West Texas are parched and carcasses of thousands of cattle are to be see in every direction In some localities no rain has fallen since last September....it is drier than it has been in twenty years. Of 7,000,000 head of cattle in Texas, one-third are in the section visited by the drouth. (The drouth came as far east as Big Springs, Texas. Cattle were dying at the rate of 900 per day. Twenty thousand carcasses are to be found. On the A and P railroad, the stench is terrific). (parenthesis in the original)

In a later article, *Hoof and Horn* (2/3/1887) reported that conditions were so bad that cattle had to be "turned loose to keep them from dying." Both range conditions and the condition of the cattle were "too poor to drive." As a result, the Aztec Company had to resort to shipping most of their cattle into Arizona by rail, because neither the cattle nor the range along their path could withstand the trip.

23. The recurring deterioration of rangeland that continually accompanied the expansion of the cattle frontier was a function of the speculative nature of the cattle industry. With large sums of money invested in herds, specific prices needed to be obtained if speculators were to recover their investment. When drought conditions prevailed and prices declined in an ensuing buyer's market, many ranchers were forced to withhold their herds from sale until prices returned to their previous levels. When a drought became protracted, as it often did in the arid West, devastation was generally wrought upon herds and ranges alike.

24. Because of the depressed market for cattle, the Aztec Company was able to purchase 22,000 head from the Pecos outfit of the Continental Cattle Company for $333,000.00 ($15 per head). This price was considerably below the $30 to $35 price that had prevailed previously. An 1887 stockholders' report indicates that the Aztec Company had spent $421,615.93 to purchase 23,000 head of cattle ($18.33 per head) (see *Hoof and Horn*:2/3/1887; Kennedy 1968:7).

25. In addition to the problems presented by weakened cattle and deteriorated transit ranges (see footnote 22), the Aztec Company had to contend with local ranchers who tried to block their arrival. Arizona ranchers, fearing the effects of overgrazing and Texas tick fever, tried unsuccessfully to have a quarantine imposed on Texas cattle by the Arizona Territorial Legislature (see *Hoof and Horn*, November, 1885 through November, 1886; Kennedy 1968:9-11).

26. Estimates of the Exact amount of cattle imported by the Aztec Company vary. While Kennedy (1968:1) claims that 34,000 head were shipped into northern Arizona, Wayte (1960:133), Peterson (1973:169) and LeVine (1977:32) report that 40,000 were imported

by the company. See *Hoof and Horn* (May, 1886 through February, 1887) for articles at least partially chronicling the arrival of Aztec Company cattle.

27. The Hashknife name derived from the brand that the Aztec Company used to mark their cattle, which resembled the kitchen tool of the same name.

28. During June of 1886, and likely on other dates as well, the Aztec Company had a letter published in various local papers in the Little Colorado River Basin reiterating the company's rights to the land it had purchased and warning all of the consequences that would befall those who trespassed on Aztec Company land (cf. *Apache County Critic*:6/3/1886). The company also failed to deal "liberally" with Mormon settlers who discovered that the land they had homesteaded for nearly a decade had been granted to the railroad and subsequently sold to the Aztec Company (see footnotes 37 and 41 below).

29. See the *Apache County Critic* between May and August of 1886 for an account of several cases of trespassing brought to trial by the Aztec Company.

30. The conflict between cattle and sheep interests in the basin resulted from the overcrowding of ranges subsequent to the arrival of Aztec Company herds. By excluding all competitors from such a large portion of the region, the Aztec Company imposed a considerable hardship on those ranchers and herders who depended upon this formerly open range. As events surrounding the Pleasant Valley War suggest (see Barnes 1931, 1932; Forrest 1952), the competition for grazing land precipitated by the arrival of the Aztec Company had an impact beyond the immediate boundaries of the Little Colorado River Basin. In addition to numerous cattle, approximately 100,000 head of sheep were regularly herded in the basin prior to the arrival of the Aztec Company, mostly by Hispanic settlers. Stocking the ranges with up to 60,000 head of cattle and maintaining exclusive access to 2 million acres imposed severe difficulties upon these other stockmen and triggered confrontations on available lands in proximity to the region. In addition, a drought in New Mexico (1885-1886) produced a difficult situation for Hispanic sheepmen in that area, forcing them to enter northern Arizona in search of grass for their herds. As one source at the time maintained, "before the coming of the railroad, there wasn't any trouble between sheep and cattlemen because there was altogether more land for everybody than they could possibly use" (see Kennedy 1968:15).

31. The Aztec Company did not participate in this feud, but several of its cowboys did, siding with the cattle interests in that basin. This Pleasant Valley is not the same as that valley in which the Mormon dairy was located.

32. During the summer of 1895 cattle sold for only $5.00 at Holbrook (Wayte 1960:139).

33. See the discussion of the drought of the 1890's (Chapter 2) and the section on overgrazing (Chapter 4) for additional information regarding the extent of livestock losses and range deterioration produced by Aztec Company livestock management.

34. The violence and banditry attributed to the cowboys of the Hashknife outfit is amply discussed in several local historical sources (cf. Barnes 1931, 1932; LeSueur n.d.; Fish n.d.a.; Forrest 1952; Wayte 1960; Kennedy 1968; and Peterson 1973), and has even served as a recent source of local historical pride. Gun battles, "shoot outs", cattle rustling and all of the other characteristics generally associated with western "cow towns" during the late nineteenth century were focused on Holbrook and suffered by the entire region. Fish (n.d.a.:71-75) chronicled approximately 50 persons killed during the years immediately following the arrival of the Aztec Company, several of whom were Mormons. Many persons suffered dearly from the widespread rustling committed by the Hashknife cowboys, including the stockholders in the Aztec Company itself (see Kennedy 1968:18-19).

35. Fish (n.d.a:23) reported that several men from Snowflake had managed to recover 28 stolen horses during May of the same year.

36. The settlers at St Joseph and St. Johns, as it turned out, did not have to negotiate with either the railroad or the Aztec Company in order to gain title to their land. St. Johns, whose claim rested on an earlier claim established by Solomon Barth, was able to circumvent the problem by incorporating under federal and territorial townsite laws (Peterson 1973:172). The settlers at St. Joseph were situated on land set aside for education and were eventually able to receive full title to their land at only nominal fees charged by the state (*ibid.*:175).

37. At $4.50 per acre, the settlers at Snowflake and Taylor were charged $20,160.00 for land which cost the Aztec Company $2,240.00.

38. At $8.00 per acre, the settlers at Woodruff were charged $5,120.00 for land granted to the railroad for free. The railroad company obviously exploited the fact that it had the settlers at Woodruff in a poor bargaining position, as $8.00 per acre was a particularly high price for land in this basin, even given the speculative climate of the time.

39. These prices were only assessed on land under cultivation, which amounted to about 1,000 acres (Fish n.d.a.:66; Peterson 1973:174).

40. Fish was referring here to the assessments rendered for maintaining the irrigation system which served these two settlements (see Chapter 5).

41. The Aztec Company demanded that the full interest accruing for the four-year term of the loan be paid, even though the debt was canceled two years ahead of schedule (Peterson 1973:174).

42. Peterson's comment was made in reference to the settlers at Snowflake and Taylor, who were freed from an obligation of $19,065 to the church. A comparable proportion of the debt owed by the settlers at Woodruff was also canceled and served as yet another of the many subsidies that the church provided to the survival of this small settlement.

43. The Aztec Company managed to retain extensive rangeland in the Little Colorado River Basin, which it still leases to local ranchers.

44. Apache County encompassed the entire region under consideration until this county was divided in half to form Navajo County in 1895.

45. For an excellent discussion of the political affairs dominating Apache County during the late nineteenth century, particularly as these affected Mormon settlers in the region, see Peterson (1973:217-241).

46. Udall (n.d.:82-83) claims that the Hispanic population at St. Johns eventually forfeited its share of the water rights at St. Johns through non-use. Udall (n.d.:114-115) also claims that Mormon-Hispanic relations improved with time. It may be that enhanced water supply, the steady decline of Mormons at St. Johns after 1885, and the possible movement of Hispanics out of farming diminished the basis of the conflict between these two groups.

Regarding the movement of Hispanics out of farming, the 1900 Census reveals that as of this date farming had become a relatively unimportant occupation among the Hispanic population at St. Johns. For the 128 Hispanics declaring an occupation at St. Johns on the 1900 Census, the following categories were the most heavily represented: sheepherders (50), farm or day laborers (28), farmers (10) and sheepraisers (8). Of the 10 persons who declared farming as their principal occupation, 7 were over 50 years of age, indicating little entrance of Hispanics into farming at St. Johns subsequent to

Mormon immigration there. A decline in the number of Hispanic farmers (perhaps as a result of the competition for suitable farm land presented by the Mormon population), combined with the existence of few independent stockmen (only 8 sheepraisers compared to 50 sheepherders) and the large number of persons employed as laborers, suggests that an economic disenfranchisement of this indigenous population may have occurred with the economic development of the region during the late nineteenth century. While it is unlikely that the Hispanic population at St. Johns experienced full economic autonomy prior to the 1880's, a decline in the overall economic (and thus political) status of this population would account, in part, for the lessened overt antagonisms expressed by its members towards others, Mormons included, who might have provided necessary employment.

47. Mormon political problems within the Little Colorado River Basin began when Apache County was formed from the eastern portion of Yavapai County and elections were held in 1879 to establish a county seat and to select public officials. Several local interests desired establishing a governmental apparatus within the region, and a contest ensued to gain early control of this newly-established organization. Mormon-non-Mormon relations deteriorated steadily with each successive election as the Ring, which gained complete control over local politics, consolidated its power and became increasingly able to employ blatant methods to enforce its own interests.

48. Part of the potential threat which the region's Mormon population presented to the interests of the Ring stemmed from the widespread non-Mormon belief that individual Mormons would vote as a unit in accordance with the dictates of their leaders and in favor of Mormon candidates whenever they ran for office. As a unified voting block, the Mormon population would have presented a formidable threat. While the size of the Mormon population at St. Johns may have been declining after 1885, the same could not be said for the basin as a whole. Furthermore, during the formative years between 1879 and 1885, when the Ring emerged and consolidated its power, the Mormon population was growing steadily almost everywhere in the basin, most notably along the upper reaches of the Little Colorado River where the Ring was centered.

49. The Ring received legal support for its program to disenfranchise the Mormon population of Apache County. In 1885, a bill was passed by the Territorial Legislature and signed into law which barred Mormons from voting in Arizona.

50. Originally, the resources of the Ring were not directed exclusively against Mormons, but were employed to undermine all opposing interests in the basin. However, as the power of the Ring increased, many other interests merged with it, leaving the Mormons as the principal and most conspicuous outside threat.

51. Udall (n.d..:92) describes one incident that occurred at St. Johns.

> "Outsiders" decided to "jump" one of our vacant city lots. They tore down the "Mormon fence" around it, and then attempted to move a small lumber house on to the premises. In no time Mormons, Mexicans, Jews and "Gentiles" assembled on the spot and feelings ran riot. Guns were flourished in the air by outsiders and it was a miracle that no lives were lost.

Referring to another incident at St. Johns, in which the Mormon patriarch Nathan Tenney was killed, Fish (n.d.a.:53) recorded the following comment:

Bullets by the dozens flew through the streets and the portion of the town where the saints resided endangering the lives of its citizens. This promiscuous shooting is only a repetition of what has transpired many times.

During the incident mentioned by Udall, he and several other Mormons who apparently went to the scene of the disturbance in order to quell a riot were arrested and taken before U. S. Court Commissioner George A. McCarter, a bitter opponent of the Mormons and publisher of the most vehement anti-Mormon newspaper in the basin. Only after later consideration by a Grand Jury were the cases against these men dropped.

52. Due to problems with the non-Mormon population at Springerville, Mormon settlers in Round Valley had to abandon their settlements near that town and consolidate into a single ward at the present location of Eagar in 1886.

53. Many of the depredations carried out by the Hashknife outfit against Mormon settlers were conducted with the full knowledge that such actions benefitted from de facto legal impunity.

54. Peterson (1973:170-171) presents several accounts of violence directed against Mormons which clearly illustrate the uncertain political atmosphere that pervaded their existence in the basin during the 1880's.

55. Due to their particularly vulnerable position, Mormons at St. Johns and in Round Valley strongly opposed the formation of Navajo County, because it threatened to leave them politically isolated from other Mormons to the west and numerically inferior to the opposing elements that would remain in Apache County (see Jensen n.d.d.:3/4/1895).

56. Quite likely, the Mormon desire for exclusiveness and the thinly-veiled contempt expressed by certain church leaders towards non-Mormons generally--as reflected in Fish's comments regarding the school issue--no doubt alienated many non-Mormons from rallying to the colonists' defense and openly invited aggression from others. The various stake minutes are filled with disparaging references to "Babylon" and the contemptible lifestyle of its inhabitants. During one such tirade in which he exhorted the faithful to become "self-sustaining and independent of babylon" and to let "these outside stores put up in our country alone", Lot Smith, President of the Little Colorado Stake, "spoke of the jealousy of those who are not of us, yet the hand of the Lord is over us, and the Kingdom of God presents something formidable to the wicked", essentially equating "non-Mormons" with "the wicked" (see Little Colorado Stake:8/2/1879). It is significant that Lot Smith's speech occurred considerably before Mormon-non-Mormon hostilities in the basin had heightened.

57. In April of 1884, a party of 100 families called to St. Johns had arrived at their destination. By the end of that year, however, the majority had left, due largely to the tense political climate that prevailed.

58. Several aggressively anti-Mormon newspapers existed in Apache County during the 1880's, including the *Arizona Pioneer*, the *Apache Chief* and the *Arizona Weekly Journal*. Perhaps the most menacing of these was the *Apache Chief*. Unambiguously sub-titled "An Anti-Mormon Newspaper", the *Apache Chief* was published by George McCarter, the U. S. Court Commissioner referred to earlier in footnote 51. See Peterson (1973:228-230) for a discussion of these newspapers.

59. The year 1885 was a particularly difficult one for Mormons in the Little Colorado River Basin. During this year, Mormon settlers had to contend with both the judicial raid and the influx of the Aztec Land and Cattle Company. In addition, 1885 represented the

zenith of the hostilities perpetrated by the St. Johns Ring and of the problems experienced by Mormon settlers at St. Johns.

60. One early pioneer recalled the following incident which, with a sense of humor, reveals the church's attitude toward the Little Colorado settlements at the time (quoted in Peterson 1973:188-189; Leone 1979:98; see also footnote 1 above):

> The church authorities were practical and philosophical. For example: The people on the Little Colorado were having a hard time, especially at Woodruff and Joseph City. They just couldn't keep those dams in, and a couple of Apostles came down to look things over and give encouragement and aid to those people. Those Apostles came on up into the more prosperous communities and at meetings asked us to donate. Quote: "Brothers and Sisters: You know the extreme difficulties our people are having along the Little Colorado. We must not let those settlements be broken up. You more prosperous people must help them. Donate your cattle, horses, grain, wagons--anything you can spare that they can use.... Listen, (and the speaker lowered his voice) If we can just keep those old people there till they die off and the young ones grow up, it will be home to those young people. They will know no other home nor want any other home. And when the dam goes out they will be just like a bunch of beavers. They won't know anything else but to go and put it in again. They will be permanently located--rooted into the soil...." Good philosophy, no? Who else would have thought of that but a Mormon colonization promotor? And, the people of the more prosperous communities contributed liberally.

61. Given the failure of the American Colonization Company's attempt to settle the Little Colorado River Basin and the fact that no substantial non-Mormon farming settlement existed in the region, it is quite likely that such settlements would not have been established were it not for the concerted Mormon effort.
62. See footnote 59.
63. Those settlements most removed from direct competition with the Aztec Company --such as Eagar and Alpine--were also situated amid productive mountain pastures. They, therefore, did not display the marked decline in livestock productivity experienced by other Mormon settlements in the basin during the 1890's (see Table 2.5).
64. As indicated above, farmers in the lower valley were converting their fields to alfalfa production during the late 1890's. A transition apparently began at this time from traditional Mormon mixed farming to livestock production on irrigated pasture. This transition is reflected in the simultaneous decline in tithing by fieldcrops and increase in tithing by livestock at St. Joseph after 1896 (see Tables 2.4 and 2.5).
65. Gates (1960:383-386) indicates that the instability characteristic of American frontier settlements was largely absent in those territories settled by Mormon pioneers and under the jurisdiction of the Mormon Church (see also Arrington 1958:411-412).
66. This is true of multi-species communities as well (cf. Elton 1958; Margalef 1968; E. Odum 1971).

CHAPTER 8

Conclusion

It is clear from the previous chapters that the environmental consequences for Mormon colonization of the Little Colorado River Basin were both direct and pervasive. However, it remains to be shown that both successful Mormon colonization of this region and specific empirical developments associated with the settlement process can be accounted for by general ecological theory. The various environmental themes of the previous chapters may now be integrated using the ecological model presented in Chapter 3 as the theoretical basis for a systematic explanation of successful Mormon colonization of the region. Specifically, this chapter will apply ecological theory to explain: (1) local and subregional differences in community development, (2) the role of resource redistribution in successful colonization, (3) the differential success of Mormon efforts at multihabitat resource redistribution, and (4) the impact of external factors on the settlement process. The chapter will conclude with a discussion of the implications raised by the successful application of ecological theory to explain Mormon colonization of the Little Colorado River Basin. In particular, implications will be drawn regarding: (1) the priority of material considerations in understanding successful Mormon colonization, (2) the explicit application of general ecological concepts and principles to human populations, and (3) the ecological basis for the evolution of complex human communities.

Ecology and the Little Colorado River Basin

As discussed in Chapter 3, the evolution of complex ecological communities is an adaptive organizational process whereby a growing population responds to changing conditions of resource availability in its environment. Since an increase in functional diversity within a specific community depends on the viability of individuals exploiting increasingly specialized and, thus, more marginal resources, greater diversity evolves as a product of increasing population size and productivity. However, due to the negative effect that maintenance costs have on energy flow in ecological systems, increasing diversity also depends on increasing net or per capita productivity. Consequently, the greatest diversity is achieved in ecological communities which simultaneously maximize aggregate productivity, per capita productivity and population size.

Continued, simultaneous increases in aggregate productivity, per capita productivity and population size within a specific community result primarily from the elimination of restrictions on the abundance and distribution of resources within a particular environment. The most significant limiting factors inhibiting the evolution of complex ecological communities are those which accompany an unproductive and variable environment. Besides restricting aggregate resource flows, unyielding and unstable habitats may also impose drains on a community which increase its maintenance costs and, thus, reduce its net productivity. Since increasing community diversity depends on increasing aggregate productivity, per capita productivity and population size, the evolution of more complex ecological communities requires the amelioration of those conditions which: (1) limit the size and reliability of aggregate productivity and/or (2) reduce available net productivity by increasing maintenance costs.

The Little Colorado River Basin presented several formidable obstacles to early Mormon settlers. Insufficient and unreliable precipitation, inadequate and highly variable growing seasons, poor soils, variable and poor-quality irrigation water, and several other specific physical features of the basin imposed drains on local populations and limits on community productivity. Although some limitations proved surmountable, others did not. The extreme aridity of much of the basin confined all significant agricultural operations to river valleys where irrigation could be practiced. Variable growing seasons further restricted such operations to river valleys at lower elevations. At the same time, the unique combination of subsidies and drains associated with specific habitats gave rise to a differential distribution of conditions conducive to the evolution of complex agricultural communities. While the conditions which prevailed within specific settlement sites differed from one another, the discontinuous and dispersed distribution of towns throughout the basin produced: (1) similar patterns of community development among settlements within the same subregion, and (2) distinct developmental patterns among settlements in different subregions. Significantly, developmental differences within and between subregions are both accounted for by general ecological theory.

Highland Settlements

Although settlements in the southern highlands had sufficient water for farming, they endured the shortest and most variable growing seasons in the basin. The length of the frost-free period was generally too short to permit the cultivation of some crops and too variable to produce reliable harvests of others. Moreover, these settlements were generally situated in narrow valleys, with the higher valleys containing poor-quality soils due largely to inadequate drainage. Consequently, aggregate productivity[1] among highland settlements was consistently the lowest in the basin, as was the size of the populations inhabiting these towns. Since variation in the length of the growing season--the principal limiting factor affecting agricultural productivity in this subregion--lay beyond human control, the highland settlements also exhibited high variability in both aggregate community productivity and population size.

A relatively high proportion of non-heads of households declared farming as their primary occupation among the most highly situated settlements.[2] Such large numbers of dependents assisting household heads in farming operations suggests that a high per capita labor investment was required to sustain even the limited agricultural productivity achieved throughout this subregion. A heavy investment of labor in the face of low and highly variable agricultural returns resulted in a low per capita productivity among these towns.

In summation, low environmental productivity and stability yielded low and highly variable aggregate community productivity among highland settlements. In addition, environmental instability imposed high maintenance costs, which resulted in low net (per capita) community productivity. As predicted by general ecological theory, settlements in the southern highlands displayed the least functional diversity of any Mormon towns in the region.

Lower Valley Settlements

While length of the growing season did not pose a problem for agricultural settlements in the lower valley, the high summer temperatures, dust storms and greater evapotranspiration rates which characterize this subregion did. Moreover, their location along the northern margin of the region forced the lower valley settlements to irrigate their crops with the poorest quality water in the entire basin. The northern portion of the basin also contained the poorest soils which, already heavily alkaline, deteriorated steadily during the settlement period as a result of continuous irrigation.

Recurring, pronounced and largely unpredictable variation in the abundance of surface water further threatened what limited agricultural productivity existed in the lower valley. Their location downstream on the Little Colorado and its tributaries subjected lower valley settlements to the largest variation in surface-water availability experienced by any towns in the basin. They suffered annually with little or no water for irrigation during the early growing season. They also had to contend with excessive river flows later in the growing season that frequently inundated fields and caused extensive damage to their dams which, due to the alluvial composition of the lower Little Colorado River bed, could not be firmly implanted. In addition, owing to the lack of suitable storage sites, the lower valley settlements were unable to construct irrigation reservoirs and, thus, remained fully susceptible to the extreme variability inherent in their surface water supply.

By the turn of the century, settlements in the lower valley had endured several times the number of dam failures of any other settlements in the basin. Dams were easily the most important and expensive infrastructural investment among the early Little Colorado Mormon settlements. The timing of a dam failure, in conjunction with the required reconstruction effort, often imposed an unbearable stress on individual towns. With frequent dam failures occurring among some of the smallest populations in the region, the lower valley settlements easily bore the highest per capita maintenance costs in the region. Repeated crop failures caused by recurring dam washouts resulted in low and sometimes negative net productivity. With only limited and highly variable productivity, settlements in the lower valley contained the most unstable populations in the basin. Indeed, so great were the maintenance costs relative to productivity that only two of the six towns established in this subregion survived the settlement period, and these two prevailed only because they received repeated subsidies that sustained them through successive crises.

As if indigenous conditions were not bad enough, lower valley settlements also bore the most negative consequences of exploitative activities elsewhere in the region. Because these settlements were situated downstream from all other towns in the basin, their fields sustained the greatest damage from increasing silt concentration in streams due to overgrazing, and their dams and fields both suffered from the increased variability in surface-water flow in the Little Colorado caused by the construction of irrigation reservoirs upstream.

The lower valley settlements were located within the most demanding and variable habitats in the entire region with regard to the availability of resources essential for the support of agriculture. Sustaining a moderate yet highly variable annual productivity and suffering the drain of excessive costs accompanying the maintenance of their agricultural systems under highly unstable environmental conditions, the lower valley settlements frequently endured years of negative net productivity which caused extensive emigration from that subregion and which repeatedly necessitated the judicious application of subsidies from external sources within and without the basin. Under these conditions, the lower valley settlements did not generate much diversification into non-farming specializations[3]--indeed, in general they did not survive. As a group, they were even less successful than settlements in the southern highlands.

In general ecological terms, lower valley settlements were situated in moderately productive yet highly unstable habitats that clearly imposed the highest community maintenance costs in the basin. Moderate environmental productivity in association with low environmental stability and very high maintenance costs yielded both low and highly variable aggregate and net community productivity. As predicted by general ecological theory, lower valley settlements contained among the smallest and most variable populations in the basin and exhibited a functional diversity insignificantly greater than that achieved among settlement in the southern highlands.[4]

It is important to emphasize that, in one sense, St. Joseph and Woodruff are not strictly representative of towns in the lower valley. These two settlements were the most successful; they survived. The available data suggests that those lower valley settlements that did not survive displayed worse developmental records than either St. Joseph or Woodruff. Significantly, the lower valley settlements that did not survive were all situated downstream from St. Joseph and Woodruff and likely suffered more intensely the very conditions that drained these two towns. The environmental differences may not have been very great among settlements in such close proximity. However, the precarious position of Woodruff and St. Joseph throughout the settlement period suggests that even small differences may have proved fatal.

Intermediate Settlements

Settlements in valleys at intermediate elevations endured neither an inadequate growing season nor an unreliable water supply. In addition, their proximity to perennial streams and suitable reservoir sites spared these settlements the surface-water variability that plagued towns in the lower valley. For intermediate settlements, then, the combined stability of growing season and surface water availability was greater than that experienced by all other settlements in the basin. The intermediate settlements were also situated within relatively large valleys which contained the most fertile soils in the region. In addition, they were able to irrigate these soils with generally abundant, silt-free water. These features combined to permit high aggregate productivities capable of supporting the largest populations in the basin. At the same time, settlements at intermediate elevations had to contend with far fewer dam failures than those in the lower valley. Consequently, possessing the largest populations and sustaining comparatively few dam failures, they endured insignificant per capita maintenance costs relative to those imposed on settlements at either higher or lower elevations.[5]

From the perspective of general ecology, large valleys, good soils and abundant, good-quality surface water yielded high environmental productivity among intermediate settlements. Stable surface-water sources (made even more stable and abundant by the construction of storage reservoirs) together with reliable and sufficient growing seasons resulted in high environmental stability for precisely those conditions that severely restricted agricultural productivity elsewhere in the basin. High environmental productivity and stability combined to produce the largest and least variable aggregate productivities and population sizes of any settlements in the basin. With aggregate productivity reliably high and maintenance costs relatively low, net productivity was also comparatively high. Given both the size and stability of their aggregate productivity, net productivity and population size, intermediate settlements, as predicted by ecological theory, evolved a greater functional diversity than any other Mormon settlements in the region.

The settlements at intermediate elevations were not all equally distinct from those at higher and lower elevations regarding the functional diversity they had achieved. While Snowflake and St. Johns clearly stood apart from the other settlements, Eagar and Taylor did not. Both productivity and population size were larger at Eagar and Taylor than at settlements in the other subregions. However, they did not approach those achieved by either Snowflake or St. Johns. Accordingly, both Eagar and Taylor exhibited smaller per capita productivities than Snowflake and St. Johns and failed to achieve comparable degrees of functional diversity.

Information on Eagar during the nineteenth century is less complete than that for many other settlements in the region. Consequently, the data needed to fully account for the deviation of this settlement from the more successful intermediate towns is not available. However, enough information does exist which suggests an explanation. Eagar is located at a higher elevation than all other intermediate settlements and, therefore, experiences environmental conditions midway between those present among the other intermediate settlements and towns situated in the southern highlands. To begin with, Eagar experiences a shorter growing season than the other intermediate settlements. The average growing seasons at Snowflake and St. Johns are 10.7% and 22% longer respectively than at Eagar. In addition, like many settlements at higher elevations, Eagar is situated in a smaller valley than the other intermediate settlements, and one which contains soils which suffer from poor drainage. Its location, therefore, is not as suitable for irrigated farming. Eagar's deviation in the direction of the mountain settlements, thus, suggests a lower productivity combined with an increased per capita labor investment in agriculture relative to other intermediate towns (see Tables 2.3 and 2.7).

The most significant ecological distinction among intermediate settlements was the uniquely advantageous environmental circumstances surrounding the development of Snowflake and Taylor. Situated near the headwaters of Silver Creek, these two towns benefitted from the most fertile soils and from the highest-quality, least-variable surface water in the entire basin. Furthermore, unlike St. Johns, the settlers at Snowflake and Taylor did not experience any major dam failures. Nor did they suffer the same economic drain from conflict with local non-Mormon interests due to their more remote location. In addition, because agriculture in these two towns was integrated into a single encompassing system, the resources of the largest aggregate population in the entire river basin maintained what was one of the least troublesome irrigation system in the region. Even the faulting caused by the formation of the Holbrook Anticline merely imposed an upper limit on water storage (mitigated in part by the reliability of Silver Creek Spring) rather than a destabilizing condition increasing the cost and uncertainty of farming.

However, despite the apparently equal availability of material resources, both productivity and population size were considerably larger at Snowflake than at Taylor. As expected, Snowflake displayed significantly greater functional diversity as well. Snowflake registered a larger number of occupations in the 1900 census, both absolutely and as a ratio of the number of persons declaring an occupation. In addition, the number of businesses and business categories recorded for Snowflake in 1905-1906 was considerably larger than that for Taylor. Finally, the extent of church organization achieved at Snowflake by the turn of the century was much greater than that attained by Taylor as well.

The marked discrepancy in population size, productivity and functional diversity between Snowflake and Taylor resulted from the close proximity of these two settlements, combined with specific material advantages accruing to Snowflake. Located in the larger and more fertile of the two adjoining valleys, Snowflake quickly emerged as the dominant settlement, supporting a larger population and achieving a greater aggregate community productivity. Just as close proximity facilitated the integration of their agricultural systems, so also did this spatial propinquity serve to unite the aggregate resource flows of these two towns.

Conclusion 199

As new activities and enterprises emerged to serve both settlements, economic considerations assured their location in Snowflake, the larger of the two towns.[6] Gradually, Snowflake acquired the majority of such operations and, serving as the focal point for the largest single population in the entire region,[7] quickly emerged as the principal Mormon settlement in the Little Colorado River Basin.[8] A realistic appraisal of the processes contributing to community development in the region requires, therefore, that Snowflake and Taylor be considered a single ecological community throughout the settlement period.

The ecological implications of variations in community development may now be summarized. Local and subregional differences in community development occurred primarily because of the differential distribution of environmental constraints, and these differences conformed to expectations derived from general ecological theory. Due largely to climatic conditions that imposed insurmountable constraints on agricultural productivity, settlements in the southern highlands were the least functionally diverse towns in the region. Although they were situated in potentially more productive habitats than the small mountain valleys to the south, settlements in the lower valley suffered from highly variable environmental conditions--most notably those associated with surface flow in the Little Colorado River--which demanded the largest maintenance costs in the basin. As predicted by ecological theory, the lower valley settlements achieved a functional diversity insignificantly greater than that displayed by towns in the southern highlands.[9]

The remaining settlements, those situated in valleys at intermediate elevations, were both productive and stable, and experienced the lowest maintenance costs of any settlements in the basin. As predicted by ecological theory, intermediate settlements contained the largest and most stable populations in the region and evolved the most functionally diverse community organizations. In addition, because it was situated in the most productive and stable habitat in the entire basin (and one that imposed the lowest maintenance costs attending agriculture); and because it served as the focus for two functionally integrated settlements commanding between them the largest population and aggregate productivity in the region, Snowflake quickly emerged as the paramount Mormon settlement in the Little Colorado River Basin. Thus, both the clear developmental distinction between the intermediate settlements and those towns situated at higher and lower elevations as well as the regional pre-eminence of Snowflake comply with specific predictions derived from the ecological model presented in Chapter 3.

Table 8.1 shows the relationship of population, productivity and stability to functional diversity among individual Mormon settlements in the Little Colorado River Basin during the nineteenth century. The left side of the table ranks settlements according to the magnitude and stability of aggregate productivity, per capita productivity and population size. Each settlement is ranked according to its mean (X) and Coefficient of Variation (V) on each of these three variables. The mean of the individual rankings provides a composite rank-order, giving a measure of the relative degree to which each settlement possessed the

Table 8.1

Rank-Order of Mormon Settlements
in the Little Colorado River Basin
1887 - 1905

Settlement	Population - Productivity - Stability Rank-Order[a]							Diversity Rank-Order					
	Population		Aggregate Productivity		Per Capita Productivity		Composite Rank-Order		Occupations 1900 Census	Businesses 1905-1906	Business Categories 1905-1906	Composite Rank-Order	
	$\bar{X}$	V	$\bar{X}$	V	$\bar{X}$	V	$\bar{X}$	R-O	#	#	#	$\bar{X}$	R-O
Intermediate:													
Snowflake/Taylor	1	1	1	2	2	2	1.67	1	1	2	1	1.33	1
St. Johns	2	2	2	3	3	3	2.33	2	2	1	2	1.67	2
Eagar	3	3	3	1	5	1	2.67	3	4.5	3	3	3.50	3
Lower Valley:													
St. Joseph	6	6	4	7	1	5	4.83	4	6.5	6	4.5	5.67	6
Woodruff	5	5	5	6	4	6	5.17	5.5	3	4	4.5	3.83	4
Mountain:													
Showlow	4	4	7	5	7	4	5.17	5.5	4.5	5	6.5	5.33	5
Alpine	7	7	6	4	6	7	6.17	7	6.5	7	6.5	6.67	7

$r_s = .884 \qquad p = <.01$

SOURCES: Tables 2.3, 2.6, 2.7, 2.8 and 2.11.
[a]Rank-order by mean ($\bar{X}$) is from highest to lowest; rank-order by coefficient of variation (V) is from lowest to highest.

conditions conducive to the evolution of complex ecological communities. A similar procedure is employed on the right side of the table to arrive at a composite measure of functional diversity. Snowflake and Taylor, for reasons already discussed, have been treated as a single community in the calculations performed. As predicted by ecological theory, a strong correlation exists among individual communities in the Little Colorado River Basin between population, productivity and stability on the one hand and functional diversity on the other. If the decline in St. Joseph's diversity ranking had not been caused by its close proximity to Holbrook, a near-perfect rank-order correlation would have been achieved.

Several conclusions may be drawn from this table. *First*, the intermediate settlements consistently rank highest across the table, supporting the previous discussion which emphasized the uniqueness of this subregion for the development of agricultural communities. *Second*, as already indicated, the lower valley settlements, while ranking higher than the mountain settlements with regard to productivity, were somewhat less stable and, therefore, do not rank clearly higher than these latter towns in terms of their functional community. *Third*, the correlation achieved when comparing the two composite rank orders clearly supports the argument that the evolution of complex community organization occurred among those settlements which simultaneously maximized aggregate productivity, per capita productivity and population size, due largely to the differential constraints imposed by local environmental conditions. *Finally*, and most significantly, the high correlation is explained by general ecological theory.

Multihabitat Resource Redistribution

Although members of the original lower valley settlements attempted to offset the effects of variable environmental conditions in their subregion by exploiting resources outside the valley, their efforts were largely unsuccessful. Irrigated farming along the lower Little Colorado demanded the most labor-intensive agriculture in the entire basin. However, only a few hundred persons inhabited the lower valley during the early years, and the labor of this small population was committed to maintaining at least three separate agricultural systems. Already overextended, available manpower in these settlements was further restricted by high dependency ratios. Settlements in the lower valley simply could not spare the labor required to effectively exploit distinct habitats located at considerable distances from the river valley, especially as this manpower was needed at precisely the time that labor was required to perform necessary farming operations along the Little Colorado. Furthermore, because all the settlements which operated the conjoint enterprises were located in neighboring habitats within the same subregion, they all exhibited similar schedules of variation in agricultural productivity and were generally equally susceptible to the same natural calamities. Consequently, when a disaster occurred, such as a flood or the loss of a dam, it tended to affect all the communities equally, forcing each town to divert manpower from the conjoint enterprises simultaneously. As persistently high maintenance costs caused settlements to be abandoned and the lower valley population to decline, the conjoint enterprises and the multihabitat exploitative strategy they represented could no longer be maintained.

As Mormon immigration into the region continued and as permanent settlements became established in diverse local habitats throughout the river basin, a relatively successful multihabitat resource-flow system did eventually emerge. This later system, based on tithing redistribution, successfully integrated the productivity of a much larger aggregate population permanently situated in numerous, distinct and widely-separated habitats and was, therefore, better able to offset the destabilizing impacts caused by the variability of resource flows in any single habitat. The effectiveness of this subsequent resource-flow system was due, in large part, to the fact that it functioned through a flexible, highly responsive and centrally-directed institutional and organizational structure which facilitated resource transfers and conversions. This system integrated the entire regional Mormon population into a single economic community which generated a substantial, centrally-administered surplus. This surplus, procured at a moderate per capita cost, represented a readily available supply of resources which could be quickly applied to subsidize local needs anywhere in the basin. While quite successful on its own, the system of tithing redistribution also benefitted from the affiliation of its constituent institutions with parent organizations outside the basin whose resource flows were independent of local environmental conditions.

The differential success of early Mormon attempts to establish a multihabitat resource-flow system is also predicted by ecological theory. As already discussed, the lower valley settlements experienced great difficulty evolving functionally diverse communities within their immediate habitats. For the same reasons, moderate aggregate productivities in the face of high maintenance costs, small and highly variable populations, low net productivities and heavy susceptibility to environmental instability were also insufficient to support the degree of functional specialization and diversity required to operate even the limited multihabitat resource-flow system attempted by these early pioneers. Thus, for the reasons outlined in the general ecological model, the system of conjoint enterprises failed as a mechanism of environmental regulation despite its cooperative orientation, communal organization and explicit ethnoecological basis.

From an ecological viewpoint, the subsequent and much expanded resource-flow system differed in every important regard from the first. The population encompassed by this later system was several times greater than that situated in the lower valley. The subsequent system also incorporated several intermediate settlements, whose productivities were far in excess of anything achieved in the lower valley and upon which the success of the system ultimately depended. Increased population size, greater aggregate and per capita productivity and reduced per capita maintenance costs resulted in a much larger net productivity in the later, much expanded system. Primary dependence of individual populations on distinct and widely-separated habitats also limited the impact of environmental variability on the continuity and reliability of resource flows in the system. Thus, unlike the system of conjoint enterprises, the system of tithing redistribution contained sufficient *redundancy* to offset the negative consequences of local environmental variability (see Abruzzi 1989). Less susceptible to environmental instability, a larger and more reliable net productivity was available to support the complex functions through which the expanded resource-flow system operated. Thus, for reasons specified by the ecological model, tithing redistribution succeeded as a mechanism of environmental regulation despite the fact that its principal supporting institutions were established to solve explicitly non-ecological problems. Ecological theory, therefore, consistently explains both the failure of the system of conjoint enterprises and the success of the later system of tithing redistribution, in terms of: (1) the conditions required for their successful development and (2) their differential capability as mechanisms regulating population-resource relationships.

The Expansion of the American Frontier

Early Mormon pioneers had to contend not only with a highly variable and unpredictable natural environment, but also with several destabilizing conditions associated with the expansion of the American frontier. The intense competition that followed the local arrival of this frontier seriously threatened not just individual settlements, but the entire colonizing effort. Although the advent of the railroad provided several important subsidies, this event also precipitated a series of incidents which, lasting nearly a decade, imposed a considerable drain on the resources of the regional Mormon population and on certain towns especially. As a result of the difficulties attending the railroad's arrival, many pioneers emigrated, and the population of most Mormon settlements in the region declined. Productive activities of all kinds faltered: Mormons increasingly abandoned stockraising, and in some locations farming; numerous Mormon businesses closed; and even the church-affiliated ACMI store in Snowflake (the most profitable co-op store in the region) came near to failing. Whole settlements were threatened economically--and in some cases politically--and the complex, church-affiliated resource-flow system was placed in considerable jeopardy. However, with the help of repeated material subsidies received from church headquarters in Salt Lake City and from other Mormon sources outside the basin, the local redistributive system was able to channel enough money, manpower and produce to endangered settlements during crises to effectively counter the challenges imposed.

By 1890 the worst was over and not a single Mormon settlement had been lost as a result of the economic and political conditions engendered by the frontier's arrival. However, due to successive crises, aggregate productivity and population size among Mormon settlements throughout the region declined. In addition, increased maintenance costs, particularly at certain settlements, had so reduced per capita (and thus net) productivity among Mormon towns in the region that the resources available to support the complex resource-flow system integrating these settlements was considerably reduced. However, despite the losses endured, the spatial diversity, concentration of resources and dispersion of maintenance costs inherent in the organization of the later Mormon redistributive system eventually enabled these settlements individually and collectively to withstand the numerous and substantial drains caused by economic and political competition in exactly the same manner they had offset the instabilities imposed by the basin's variable natural environment. Likewise, church sources outside the basin provided critical subsidies when the costs imposed exceeded the resources of the local resource-flow system.

CONCLUSIONS AND IMPLICATIONS

Several conclusions may be drawn from the preceding investigation of Mormon colonization of the Little Colorado River Basin. The most direct conclusion concerns the effect of material conditions on the settlement process. Of central importance were those natural environmental conditions that acted as either subsidies to or drains upon community development in the region. The similar developmental histories of towns in the same general subregion and the developmental distinctions displayed by towns in different subregions evidences the strong formative impact of the natural environment on the settlement process. Local developmental differences within subregions resulted from environmental influences as well. The less successful highland and lower valley settlements suffered more intensely the specific natural environmental conditions that constrained community development in their respective subregions. Likewise, the emergence of Snowflake as the leading Mormon settlement was based on its greater access to the precise environmental conditions that made intermediate settlements the most successful Mormon towns in the region. Natural environmental conditions also largely determined the relative success of Mormon efforts to establish a viable multihabitat system of resource redistribution.

The natural environment was not the only set of material conditions influencing the settlement process. The competition introduced by the expansion of the American frontier imposed severe material drains on local communities and seriously undermined the complex organizational structure that integrated the Little Colorado Mormon settlements into a unified resource-flow system. By imposing a severe drain on local Mormon resources, the intense competition that followed the railroad's arrival severely hindered the development of several local communities and threatened the success of the colonization effort.

The Mormon church and local Mormon institutions performed a critical role in the settlement process primarily through the material resources they provided to subsidize the colonization effort. Their church affiliation provided Mormon settlers with an organizational structure that facilitated an effective redistribution of local resources in order to overcome the material drains imposed by environmental instability, as well as by economic and political competition. Indeed, the adaptive advantage of nineteenth century Mormon institutions under variable frontier conditions appears to stem more from material considerations

Conclusion 205

than from the social and economic consequences of cooperative Mormon values. While such values were important proximate mechanisms underlying successful Mormon colonization of the region, the success of this colonization effort and the importance of the church connection had an ultimate material basis: the judicious redistribution of surplus resources among individual settlements located in separate and independent habitats and the continued receipt of material subsidies from church headquarters in Salt Lake City and other Mormon sources outside the basin.

The preceding investigation clearly demonstrates that a precise understanding of the settlement process depends on a consideration of the specific material considerations to which local Mormon populations had to adapt and of the ecological viability of their adaptive response to variable environmental conditions. A reference to Mormon values explains neither local and subregional differences in community development nor the differential success of Mormon efforts to establish a viable system of resource redistribution. The conjoint enterprises of the lower valley settlements failed as mechanisms of environmental regulation despite their cooperative orientation and communal organization because the material circumstances associated with them were inadequate when compared to the natural environmental conditions they had to overcome. Conversely, tithing redistribution provided an ecologically viable system of environmental regulation because the material conditions of its resource flows effectively mitigated the impact of environmental instability, despite the fact that its principal institution were established for manifestly nonecological purposes and that it succeeded in part because participation in the system was individually profitable. The present research, therefore, suggests a serious reevaluation of traditional value-oriented explanations for successful Mormon colonization of the American West.

The most significant direct conclusion of this research concerns the successful application of ecological theory to explain Mormon settlement in the Little Colorado River Basin. Not only does general ecological theory provide an effective theoretical framework for explaining the overall success of the colonization effort, but the specific model outlined in Chapter 3 provides a synthetic account of empirical developments associated with the settlement process. Specifically, the model consistently and parsimoniously explains: (1) local and subregional differences in community development; (2) the role of resource redistribution in successful colonization; (3) the differential success of Mormon efforts in establishing an effective resource redistribution systems; and (4) the impact of external factors on the settlement process. The present investigation demonstrates that both local and subregional differences in community development among Little Colorado Mormon settlements conformed to expectations derived from comparative ecological research. A clear positive association existed between community productivity and stability on the one hand and community diversity on the other. In addition, the most complex community organizations occurred among settlements that were situated in the most productive and stable habitats for agriculture and that suffered the least from external exploitation. The research has also shown that successful Mormon colonization of this region resulted largely from the development of an integrated system of resource redistribution that enabled individual settlements to endure the impact of local environmental variation. Moreover, as predicted by general ecological theory, the present investigation has shown that only one of the two systems of resource redistribution attempted was successful, due largely to the increased redundancy of resource flows provided by its greater diversity.

The successful application of ecological theory to this colonization effort has several broader theoretical and methodological implications. To begin with, ecological theory provides a more systematic and precise explanation of Mormon colonization of the region than has been offered by previous researchers or than can potentially be achieved through alternate historical or anthropological explanations of community development (see Chapter 1). In addition, the present application of ecological theory explains specific local developments within a deductive and operational explanatory framework that leads to testable hypotheses rather than by merely using ecological concepts and principles as heuristic metaphors. Besides providing a more precise and systematic explanation of local empirical developments, this methodology has proven more susceptible to empirical verification and, therefore, to objective evaluation. It also offers a greater opportunity for exporting the present analysis to other ethnographic contexts.

The ecological model also systematically accounts for the destabilizing effect of external impacts on the settlement process whether those impacts derive from natural or social environmental conditions. In addition, ecological theory provides an explanation of events surrounding the settlement process that is systemic rather than reductionist. That is, the present investigation has not simply shown that environmental factors influenced community development in the region, but rather that general ecological considerations underlay the local evolution of complex human communities.

At the same time, the systematic application of ecological theory in this case enables ecological anthropology to make a substantive contribution to the development of general ecological theory rather than merely serve as a passive recipient of ecological ideas. For example, the present investigation bears directly on the debate regarding the relationship between diversity and stability in ecological systems (see Chapter 3, footnote 14). The positive association established between community productivity and stability on the one hand and community diversity on the other, together with the fact that the most complex community organization occurred among settlements situated in the most productive and stable habitats clearly supports those researchers who claim that diversity derives from stability in ecological systems, while the greater success of tithing redistribution as a mechanism of environmental regulation upholds those who propose the alternate thesis that stability derives from diversity within ecological communities. Indeed, the research discussed here suggests that these competing theses may merely reflect processes operating at different levels in hierarchically organized ecological systems (see Abruzzi 1987).

Finally, and most importantly, the successful application of ecological theory to explain historical developments in the Little Colorado River Basin affirms the utility and validity of applying general ecological principles in social analysis. It lends support to the notion that human communities represent a subset of general ecological systems and that their organization and evolution conform to the same theoretical considerations that explain the development of other ecological communities. This research, therefore, clearly supports those who propose a broader ecological perspective on the human condition and who argue for the development of a unified ecological analysis of social behavior among human and nonhuman populations.

NOTES

1. In the following explanation, tithing data is employed to represent productivity (see Chapter 3). Total tithing and per capita tithing will serve as indicators of aggregate and per capita productivity respectively (see Tables 2.3 and 2.7). Diversity will be measured by a combination of the data in Tables 2.8 and 2.11, and data on population size are contained in Table 2.6.
2. See Chapter 4, footnote 36.
3. While settlers at St. Joseph paid the highest average per capita tithing in the basin, the amount of tithing paid there was also among the most variable. Moreover, this high average was largely a function of a sharp increase in per capita tithing paid after 1896, when settlers there began adopting non-farming occupations. The variability associated with per capita tithing at St. Joseph decreased subsequent to the procurement of non-farming employment as well.
4. Environmental instability negated the developmental advantage of greater environmental productivity in the lower valley relative to the southern highlands in the same way that it did in Sanders' (1968) comparison of species diversity in stable temperate versus unstable tropical ecosystems (see Chapter 3).
5. Although St. Johns suffered several major dam failures, the largest and most devastating of these occurred in 1915, subsequent to the period under consideration.
6. This appropriation of functions by Snowflake and the consequent functional "simplification" of Taylor was similar to the situation experienced by St. Joseph following the establishment of Holbrook (see footnote 9 below).
7. The combined population of Snowflake and Taylor was, on the average, 40% larger in size than the Mormon population at St. Johns. This difference was even greater during later years. In addition, those businesses listed for St. Johns in 1905-1906 were for the whole town, Mormons and non-Mormons included, while all businesses in the Snowflake-Taylor area were Mormon-owned. The combined population of Snowflake and Taylor was even larger than that of Holbrook, the regional shipping point, which in 1900 included about 600 persons.

8. The literature regarding Mormon colonization of the Little Colorado River Basin clearly indicates the economic, political and social ascendancy of Snowflake. Indeed, in much of the literature Snowflake emerges as more nearly approximating the ideal Mormon agricultural community than any other settlement in the region This perception of Snowflake is clearly expressed in the following comments made by Charles Peterson (1967:4) in which he recalls his impressions as a youth growing up in the Little Colorado River Basin. While Peterson's comments relate to a later period, the impression presented applies equally well to the settlement years.

> As far as my early perception goes, Snowflake, Arizona, was *the* Mormon town. It lay at the center of all things. Outward from it in order of descending importance were the small valley on whose level floor it lay, the two-dozen towns and near-towns that comprised the Little Colorado Colony, and a set of regional influences that pushed or nudged at the town's character from Utah and Salt Lake City in the north, the Salt River Valley and Phoenix in the south, New Mexico and Albuquerque in the east, and somewhere out there to the west California and Los Angeles.

> Snowflake was not only the center of the world but it sat right with the world and to my youthful eyes was perfect in its wholeness. The four-square of its grid pattern was firm and fully formed. One had a sense of being comfortably within the physical unit formed by its six north-south running streets, its nine east-west running streets, and its forty blocks. By contrast other towns struck me as being out of harmony. Neighboring Taylor--whose kids seemed hard-knuckled and a little more profane than Snowflake's God-fearing young--was scattered and split by a creek, the grid of its streets deformed and its capacity to give one a sense of being encompassed incomplete. Joe City's grid was overwhelmed by the dual thrust of the Santa Fe Railroad and Highway 66, which rushed east and west along its main street. In the mountain towns to the south, the grid pattern never took at all or phased out into surrounding pines and homesteads. Mormon St. Johns, on the other hand, was only half a town, with Mexicans claiming the rest, while Mesa, in the Salt River Valley, seemed bound in blacktop and concrete, a small city with a different tempo and soul. How wonderful it was to live in the world's only perfect social unit.

9. The overlap in diversity ranking between lower valley and mountain settlements is largely the result of a deviation in the ranking of St. Joseph (see Table 8.1). St. Joseph's discrepancy derived from its close proximity to Holbrook, the regional marketing center, and its loss of productive functions to that town.

References Cited

Abruzzi, W. S. 1979. Population Pressure and Subsistence Strategies among the Mbuti Pygmies. *Human Ecology* 7: 183-189.
-----. 1980. Flux among the Mbuti Pygmies of the Ituri Forest: An Ecological Interpretation. In *Beyond the Myths of Culture: Explorations in Cultural Materialism*, edited by E. Ross, pp. 3-31. New York: Academic Press.
-----. 1982. Ecological Theory and Ethnic Differentiaton Among Human Populations. *Current Anthropology* 23:13-35.
-----. 1985. Water and Community Development in the Little Colorado River Basin. *Human Ecology* 13:241-269.
-----. 1987. Ecological Stability and Community Diversity during Mormon Colonization of the Little Colorado River Basin. *Human Ecology* 15:317-338.
-----. 1988. Ecological Concepts in Human Ecology. Paper presented at the 12th International Congress of Anthropological and Ethnological Sciences, Zagreb, Yugoslavia, July 24-31.
-----. 1989. Ecology, Resource Redistribution, and Mormon Settlement in Northeastern Arizona. *American Anthropologist* 91:642-655.
Adams, R. N. 1975. *Energy and Structure: A Theory of Social Power*. Austin: University of Texas Press.
Akers, J. P. 1964. *Geology and Ground Water in the Central Part of Apache County, Arizona*. U.S. Geological Survey Water Supply Paper 1771. U.S. Government Printing Office. Washington, D.C.
Alexander, R.D., and G. Borgia. 1978. Group Selection, Altruism, and the Levels of Organization of Life. *Annual Review of Ecology and Systematics* 9:449-474.
Apache County. n.d. Abstract of Assessment Rolls. Report of the Clerk of the Board of Supervisors, Apache County. St. Johns, Arizona.
Arrington, L.J. 1951. Taming the Turbulent Sevier: A Story of Mormon Desert Conquest. *Western Humanities Review* 5:396-406.
-----. 1954. Orderville, Utah: A Pioneer Mormon Experiment in Economic Orgnization. *Utah State Agricultural College Monograph Series*, Vol. 2, No 2.
-----. 1958. *Great Basin Kingdom: Economic History of the Latter-Day Saints, 1830-1900*. Lincoln: University of Nebraska Press.
Arrington, L.J., F.Y Fox, and D.L. May. 1976. *Building the City of God: Community and Cooperation among the Mormons*. Salt Lake City: Deseret Book Company.
Ashby, W.R. 1956. *An Introduction to Cybernetics*. London: Chapman and Hall.
Athens, J.S. 1977. Theory Building and the Study of Evolutionary Processes in Complex Societies. In *For Theory Building in Archaeology*. edited by L.R. Binford, pp.353-384. New York: Academic Press.
Babcock, H.M., and C.T. Snyder. 1947. *Ground-Water Resources of the Holbrook Area, Navajo County, Arizona*. U.S. Geological Survey, Ground-Water Resources of Arizona 4.

References Cited

Bailey, F.G. 1957. *Caste and the Economic Frontier*. Manchester: Manchester University Press.
Barth, F. 1956. Ecological Relationships of Ethnic Groups in Swat, North Pakistan. *American Anthropologist* 58:1079-1089.
-----. 1966 *Models of Social Organization*. Royal Anthropological Institute, Occasional Paper No 23.
-----. 1967. On the Study of Social Change. *American Anthropologist* 69:661-669.
-----. 1969. Introduction. In *Ethnic Groups and Boundaries*, edited by F. Barth, pp.9-38. Boston: Little, Brown.
Barash, D.P. 1977. *Sociobiology and Behavior*. New York: Elsevier.
Barth, I. 1973. How St. Johns Was Settled. *Arizona Cattlelog* (November):18-24.
Barnes, W.C. 1931. The Pleasant Valley War of 1887: Its Genesis, History, and Necrology, Part I. *Arizona Historical Review* 4 (October):5-34.
-----. 1932. The Pleasant Valley War of 1887: Its Genesis, History, and Necrology, Part II. *Arizona Historical Review* 4 (January):23-40.
Bates, D., and S. Lees. 1979. The Myth of Population Regulation. In *Evolutionary Biology and Human Social Behavior: An Anthropological Perspective*, edited by N. Chagnon and W. Irons, pp. 273-289. North Scituate, Massachusetts: Duxbury Press.
Belshaw, C. 1967. Theoretical Problems in Economic Anthropology, In *Social Organization: Essays Presented to Raymond Firth*. edited by M. Freedman, pp. 25-42. Chicago: Aldine.
Bennett, J.W. 1967. *Hutterian Brethren: The Agricultural Economy and Social Organization of a Communal People*. Stanford: Stanford University Press.
-----. 1969. *Northern Plainsmen: Adaptive Strategy and Agrarian Life*. Chicago: Aldine-Atherton.
-----. 1976. *The Ecological Transition: Cultural Anthropology and Human Adaptation*. New York: Pergamon.
Berry, M. A. 1910. The Friendly Rivals, Apache and Navajo Counties. *Arizona New State Magazine* (December):18-19.
Blau, P. 1967. *Exchange and Power in Social Life*. New York: Wiley.
Boserup, E. 1965. *The Conditions of Agricultural Growth: The Economics of Agrarian Change under Population Pressure*. Chicago: Aldine.
Brodbeck, M. 1962. Explanation, Prediction, and "Imperfect" Knowledge. In *Scientific Explanation, Space and Time*. H. Fiegl and G. Maxwell, eds. Minneapolis: University of Minnesota Press.
Brookhaven National Laboratory. 1969. Diversity and Stability in Ecological Systems. *Brookhaven Symposia in Biology* No. 22. Springfield: U.S. Department of Commerce.
Brown, L.L., and E.O. Wilson. 1956. Character Displacement. *Systematic Zoology* 5:49-64.
Bureau of the Census. 1890. *1890 Decennial Census*. Washington, D.C.: U.S. Department of Commerce.
-----. 1900. *1900 Decennial Census*. Washington, D.C.: U.S. Department of Commerce.
-----. 1930. *1930 Decennial Census*. Washington, D.C.: U.S. Department of Commerce.
Bureau of Reclamation. 1947. *Snowflake Project Arizona*. Project Planning Report 3-8b.2-0. Washington, D.C.: U.S. Department of the Interior.

References Cited

-----. 1950. *Report on Joseph City Unit, Holbrook Project, Arizona*. Project Planning Report 3-8b.6-1. Washington, D.C.: U.S. Department of the Interior.
Bushman, J. n.d. Appraising Book and Company Business (1880-1915). Historical Department, Church of Jesus Christ of Latter-Day Saints. Salt Lake City.
Campbell, D.T. 1965. Variation and Selective Retention in Socio-Cultural Evolution. In *Social Change in Developing Areas*. H.R. Barringer, et.al., eds. Pp. 19-49. Cambridge: Schenkmen.
Carniero, R.L. 1962. Scale Analysis as an Instrument for the Study of Cultural Evolution. *Southwestern Journal of Anthropology* 18:149-169.
-----. 1967. On the Relationship between Size of Population and Complexity of Social Organization. *Southwestern Journal of Anthropology* 23:234-243.
-----. 1968. Ascertaining, Testing and Interpreting Sequences of Cultural Development. *Southwestern Journal of Anthropology* 24:354-374.
-----. 1970. A Theory of the Origin of the State. *Science* 169:733-738.
Carniero, R.L., and S.F. Tobias. 1963. The Application of Scale Analysis to the Study of Cultural Evolution. *Transactions of the New York Academy of Science* 25:196--207.
Chagnon, N.A., and W. Irons, eds. 1979. *Evolutionary Biology and Human Social Behavior: An Anthropological Perspective*. North Scituate, Massachusetts: Duxbury Press.
Chayanov, A.V. 1966. *The Theory of Peasant Economy*. Homeword, Illinois: American Economic Association.
Clark, P.J., P.T. Eckstrom, and L.D. Linden. 1964. On the Number of Individuals Per Occupation in a Human Society. *Ecology* 45:367-372.
Cody, M.L., and J.M. Diamond, eds. 1975. *Ecology and Evolution of Communities*. Cambridge: Belknap Press.
Cohen, Mark N. 1977. *The Food Crisis in Prehistory: Overpopulation and the Origins of Agriculture*. New Haven: Yale University Press.
Collingswood, R.G. 1974. Human Nature and Human History. In *The Philosophy of History*, edited by P. Gardiner, pp.17-40. London: Oxford University Press.
Collins, P.W. 1964. *The Logic of Functional Analysis in Anthropology*. Unpublished PhD. Dissertation, Department of Philosophy, Columbia University.
Colorado River Commission of Arizona. 1940. *Arizona Stream Flow Summary*. Phoenix.
Colton, H.S. 1937. Some Notes on the Original Condition of the Little Colorado River: A Side Light on the Problem of erosion. *Museum Notes of the Museum of Northern Arizona* 10(No. 6):17-20.
Committee of Settlement of Affairs of Sunset United Order. 1886. Minutes of Committee of Setlement of the Affairs of the Sunset United Order held at Woodruff August 10th, 1886. Box 8a, Folder 8, Special Collections, Northern Arizona University Library Flagstaff.
Coontz, S.H. 1961. *Population Theories and the Economic Interpretation*. London: Routledge and Kegan Paul.
Crombie, A.C. 1947. Interspecific Competition. *Journal of Animal Ecology* 16:44-73.
Culbertson, J. 1971. *Economic Development: An Ecological Approach*. New York: Alfred A. Knopf.
Dames & Moore, Inc. 1973. *Environmental Report, Cholla Power Project, Joseph City, Arizona*. Phoenix: Arizona Public Service Company.

References Cited

Day, R.H, and T. Grove, eds. 1975. *Adaptive Economic Models*. New York: Academic Press.

Dray, W. 1974. The Historical Explanation of Actions Reconsidered. In *The Philosophy of History*, edited by P. Gardiner, pp.66-89. London: Oxford University Press.

Duncan, H.S. 1972. Paired Communities and Differential Development: A Study of Four Settlements in Northeast Arizona. Unpublished M.A. Thesis, Department of Anthropology, University of California at Los Angeles.

Eastern Arizona Stake. n.d. Minutes of the Eastern Arizona Stake Conferences. Historical Department, Church of Jesus Christ of Latter-Day Saints. Salt Lake City.

Ehrlich, P.R, and LC. Birch. 1967. The "Balance of Nature" and "Population Control". *The American Naturalist* 101:97-107.

Elton, C.S. 1958. *The Ecology of Invasions by Animals and Plants*. London: Methuen and Company.

Engelberg, J., and L.L. Boyarsky. 1979. The Noncybernetic Nature of Ecosystems. *American Naturalist* 114:317-324.

Enz, RW, and M. Weller. 1964. *Summary of Snow Survey Measurements for Arizona, 1938-1964 Inclusive*. Portland: U.S. Department of Agriculture, Soil Conservation Service.

Faris, J.C. 1975. Social Evolution, Population, and Production. In *Population, Ecology and Social Evolution*, edited by S. Polgar, pp. 235-271. The Hague: Mouton.

Fidler, R.C. n.d. Urbanism as Ecological Adaptation. In *Urban Population of Southeast Asia*, edited by G.H. Krausse, _____

Fish, J. n.d.a. History of the Eastern Arizona Stake of Zion and the Establishment of the Snowflake Stake. Historical Department, Church of Jesus Christ of Latter-Day Saints. Salt Lake City.

-----. n.d.b. Autobiography. Historical Department. Church of Jesus Christ of Latter-Day Saints. Salt Lake City.

Flannery, K.V. 1972. The Cultural Evolution of Civilizations. *Annual Review of Ecology and Systematics* 1:399-426.

Forrest, E.R. 1952. *Arizona's Dark and Bloody Ground, Revised Edition*. Caldwell, Idaho: Caxton Printers.

Foster, G.M. 1967. *Tzintzuntzan: Mexican Peasants in a Changing World*. Boston: Little, Brown.

Frank, A.G. 1967. *Capitalism and Underdevelopment in Latin America: Historical Studies of Chile and Brazil*. New York: Monthly Review Press.

Fried, M. 1967. *Evolution of Political Society: An Essay in Political Anthropology*. New York: Random House.

Friedman, J. 1979. Hegelian Ecology: Between Rousseau and the World Spirit. In *Social and Ecological Systems*, edited by P.C. Burnham and R.F. Ellen, pp. 253-270. London: Academic Press.

Gall, P.L., and A. Saxe. 1977. The Ecological Evolution of Culture: The State as Predator in Prehistory. In *Exchange Systems in Prehistory*, edited by T. Earle and J.E. Ericson, pp. 255-268. New York: Academic Press.

Gates, P.W. 1960. *The Farmer's Age: Agriculture, 1815-1860*. New York: Holt, Rinehart and Winston.

Gause, G.F. 1934. *The Struggle for Existence*. Baltimore: Williams and Wilkins.

Gazetteer Publishing Company. 1905. *Arizona Business Directory, 1905-1906*. Denver.

Gelfand, D.E., and R.D. Lee eds. 1973. *Ethnic Conflicts and Power: A Cross-National Perspective*. New York: Wiley.
Gibson, L.J., and R.W. Reeves. 1970. Functional Bases of Small Towns: A Study of Arizona Settlements. *Arizona Review* 19:19-26.
Gookin, W.S., et.al. 1972. *The Comprehensive Plan 1990 for Navajo County, Arizona*. Scottsdale, Arizona.
Graves, T.D., N.B. Graves, and M.J. Kobrin. 1969. Historical Inferences from Guttman Scales: The Return of Age Area Magic? *Current Anthropology* 10:317-338.
Greenwood, N.H. 1960. *A Geographical Survey of the Upper Watershed of the Little Colorado River, Arizona*. M.S. Thesis, Department of Geography, Brigham Young University.
Guild, W.E. 1891. *Arizona: Its Commercial, Industrial and Transportation Interests*. Bureau of Statistics, Treasury Department. Washington, D.C.: U.S. Government Printing Office.
Hagen, E. 1962. *On the Theory of Social Change*. New York: Dorsey Press.
Haggett, P. 1966. *Locational Analysis in Human Geography*. London: Halstead.
Hairston, G., F.E. Smith, and L.B. Slobodkin. 1960. Community Structure, Population Control, and Competition. *The American Naturalist* 94:421-425.
Hardin, G. 1960. The Competitive Exclusion Principle. *Science* 131:1292-1297.
Hardy, B. 1969. The Trek South: How the Mormons Went to Mexico. *Southwestern Historical Quarterly* 73:197-210.
Harrell, M.A., and E.B. Eckel. 1939. *Ground-Water Resources of the Holbrook Region, Arizona*. U.S. Geological Survey Water-Supply Paper 836-B. Washington, D.C: Government Printing Office.
Harris, M. 1959. The Economy Has No Surplus? *American Anthropologist* 61:185-199
-----. 1964. *Patterns of Race in the Americas*. New York: Walker.
-----. 1968. *The Rise of Anthropological Theory*. New York: Thomas Y. Crowell.
-----. 1977. *Cannibals and Kings: The Origins of Culture*. New York: Random House.
-----. 1979. *Cultural Materialism: The Struggle for a Science of Culture*. New York: Random House.
-----. 1980. *Culture, People, Nature, Third Edition*. New York: Harper and Row.
Hempel, C. 1942. The Function of General Laws in History. *The Journal of Philosophy* 39:35-48.
-----. 1965. *Aspects of Scientific Explanation: And Other Essays in the Philosophy of Science*. New York: Free Press.
Historical Department, Church of Jesus Christ of Latter-Day Saints. n.d. *Statistical Reports of the Church of Jesus Christ of Latter-Day Saints*. Salt Lake City.
Hogbin, H.I. 1964. *A Guadalcanal Society: The Kaoka Speakers*. New York: Holt, Rinehart and Winston.
Hornocker, M. 1970. *An Analysis of Mountain Lion Predation Upon Mule Deer and Elk in the Idaho Primitive Area*. Wildlife Monograph 22, The Wildlife Society.
Jensen, A. n.d.a. Little Colorado Stake Manuscrript History. Historical Department, Church of Jesus Christ of Latter-Day Saints. Salt Lake City.
-----. n.d.b. Eastern Arizona Stake Manuscript History. Historical Department, Church of Jesus Christ of Latter-Day Saints. Salt Lake City.
-----. n.d.c. Snowflake Stake Manuscript History. Historical Department, Church of Jesus Christ of Latter-Day Saints. Salt Lake City.

References Cited

-----. n.d.d. St. Johns Stake Manuscript History. Historical Department, Church of Jesus Christ of Latter-Day Saints. Salt Lake City.

Jorgenson, J.C. 1971. *Indians and the Metropolis*. In The American Indian and Urban Society. J.O. Wadell and O.M. Watson, eds. Pp. 67-113. Boston: Little, Brown.

Jurwitz, L.R. 1954. *Rainfall in Arizona*. Arizona Highways (July):8-15.

Kennedy, S.A. 1968. A General History of the Hashknife Range under the Aztec Land and Cattle Company, Limited. Unpublished Manuscript. Arizona Collection, Arizona State University Library. Tempe.

Kester, G., et.al. 1964. *Soil Survey: Holbrook-Showlow Area, Arizona*. U.S. Department of Agriculture, Soil Conservation Service. Washington, D.C.: U.S. Government Printing Office.

Kottak, C.P. 1982. *The Past in the Present: History, Ecology, and Cultural Variation in Highland Madagascar*. Ann Arbor: University of Michigan Press.

Kummer, H. 1971. *Primate Societies: Group Techniques of Ecological Adaptation*. Chicago: Aldine.

Kunkel, J. 1970. *Society and Economic Growth: A Behavioral Perspective of Social Change*. London: Oxford University Press.

LaRue, E.C. 1916. *Colorado River and Its Utilization*. U.S. Geological Survey Water-- Supply Paper 395. Washington, D.C.: U.S. Government Printing Office.

Lee, R.B. 1968. What Hunters Do for a Living, or, How to Make Out on Scarce Resources. In *Man the Hunter*, edited by R.B. Lee and I. DeVore, pp. 30-48. Chicago: Aldine.

-----. 1972. !Kung Spatial Organization: An Ecological and Historical Perspective. *Human Ecology* 1:125-147.

Lees, S., and D. Bates. 1984. Environmental Events and the Ecology of Cummulative Change. In *The Ecosystem Concept in Anthropology*, edited by E. Moran, pp. 133-159. Boulder: Westview Press.

Leigh, E.G., Jr. 1975. Population Fluctuations, Community Stability, and Environmental Variability. In *Ecology and Evolution of Communities*, edited by M.L. Cody and J.M. Diamond, pp. 51-73. Cambridge: Belknap Press.

-----. 1977. How Does Selection Reconcile Individual Advantage with the Good of the group? *Proceedings of the National Academy of Sciences* 74:4542-4546.

Leone, M.P. 1972. The Evolution of Mormon Culture in Eastern Arizona. *Utah Historical Quarterly* 40:122-141.

-----. 1974. The Economic Basis for the Evolution of Mormon Religion. In *Religious Movements in Contemporary America*, edited by I.I. Zaretsky and M. Leone, pp. ____. Princeton: Princeton University Press.

-----. 1979. *The Roots of Modern Mormonism*. Cambridge: Harvard University Press.

LeSueur, J.W. n.d. Trouble with the Hashknife Company. James W. LeSueur Collection, File LeSu 6445, Arizona Pioneer Historical Society, Tucson.

LeVine, A.J. 1977. *From Indian Trails to Jet Trails: Snowflake's Centennial History*. Snowflake, Arizona: Snowflake Historical Society.

LeVine, R.A. 1973. Research Design in Anthropological Fieldwork. In *A Handbook of Method in Cultural Anthropology*, edited by R. Naroll and R. Cohen, pp. 183-195. New York: Columbia University Press.

Levins, R. 1968. *Evolution in Changing Environments*. Princeton: Princeton University.

References Cited

Lewis, O. 1966. *LaVida: A Puerto Rican Family in the Culture of Poverty, San Juan and New York*. New York: Vintage Press.
Lewontin, R.C., ed. 1968. *Population Biology and Evolution*. Syracuse: Syracuse University Press.
Lightfoot, K. 1980. Mormon Sociopolitical Development in Northern Arizona, 1876-1906: Implications for a Model of Prehistoric Change. *Ethnohistory* 27:197-223.
Little Colorado River Plateau Resource Conservation and Development Project. 1971. *Little Colorado River Plateau Resource Conservation and Development Project Plan*. Holbrook, Arizona.
Little Colorado Stake. n.d. Minutes of the Little Colorado Stake Conferences. Historical Department, Church of Jesus Christ of Latter-Day Saints. Salt Lake City.
Little, M.A.., and G.E.B. Morren, Jr. 1976. *Ecology, Energetics, and Human Variability*. Dubuque, Iowa: Wm. C. Brown.
MacArthur, R.H. 1955. Fluctuations of Animal Populations and a Measure of Community Stability. *Ecology* 36:533-537.
-----. 1972. *Geographical Ecology*. New York: Harper and Row.
MacArthur, R., and J. MacArthur. 1961. On Bird Species Diversity. *Ecology* 42:594-598.
Mann, L.J. 1976. *Ground-Water Resources and Water Use in Southern Navajo County, Arizona*. Arizona Water Commission Bulletin 10, Phoenix.
Margalef, R. 1968. *Perspectives in Ecological Theory*. Chicago: University of Chicago Press.
Mauss, M. 1954. *The Gift*. New York: Free Press (original 1924).
May, R.M. 1973. *Stability and Complexity in Model Ecosystems*. Princeton: Princeton University Press.
Mayr, E. 1963. *Animal Species and Evolution*. Cambridge: Belknap Press.
McClelland, D.C. 1961. *The Achieving Society*. Princeton: Van Nostrand and Company.
McClintock, J.H. 1921. *Mormon Settlement in Arizona: A Record of Peaceful Conquest of the Desert*. Phoenix: Manufacturing Stationers, Inc.
McKenny, 1882. *McKenny's Business Directory of Central and Southern California, Arizona, New Mexico, Southern Colorado and Kansas*. Arizona Department of Library and Archives.
Meggers, B.J. 1971. *Amazonia: Man and Culture in a Counterfeit Pardise*. Chicago: Aldine-Atherton.
Meining, D.W. 1965. The Mormon Culture Region: Strategies and Patterns in the Geography of the American West, 1847-1964. *Annals of the Association of American Geographers* 55:191-220.
-----. 1971. *Southwest: Three Peoples in Geographical Change, 1600-1970*. London: Oxford University Press.
Miller, M.L., and K. Larsen. 1975. *Soil Survey of Apache County, Arizona: Central Part*. U.S. Department of Agriculture, Soil Conservation Service. Washington, D.C.: U.S. Government Printing Office.
Moran, E. ed. 1984. *The Ecosystem Concept in Anthropology*. Boulder: Westview Press.
Morgan, L.H. 1877. *Ancient Society*. New York: Holt, Rinehart and Winston.
Moynihan, D.P. 1965. *The Negro Family: The Case for National Action*. Washington, D.C.: Office of Policy Planning and Research, U.S. Department of Labor.
Murdock, G.P. 1949. *Social Structure*. New York: Macmillan.
-----. 1957. World Ethnographic Sample. *American Anthropologist* 59:664-687.

References Cited

Nag, M., B. White, and R.C. Post. 1978. An Anthropological Approach to the Study of the Economic Value of Children in Java and Nepal. *Current Anthropology* 19:293-306.

Nagel, E. 1974. Determinism in History. In *The Philosophy of History*, edited by P. Gardiner, pp. 187-215. London: Oxford University Press.

-----. 1979. *The Structure of Science: Problems in the Logic of Scientific Explanation.* Hackett: Indianapolis.

Naroll, R. 1956. A Preliminary Index of Social Development. *American Anthropologist* 58:687-715.

Nelson, C. ed. 1973. *The Desert and the Sown: Nomads in the Wider Society.* Berkeley: University of California Insitute of International Studies.

Netting, R.M. 1968. *Hill Farmers of Nigeria: Cultural Ecology of the Kofyar of the Jos Plateau.* Seattle: University of Washington Press.

-----. 1977. *Cultural Ecology.* Menlo Park: Cummings.

Newcomer, PJ. 1972. The Nuer Are Dinka: An Essay on Origins and Environmental Determinism. *Man* 7:5-11.

Nichol, A.A. 1937. *The Natural Vegetation of Arizona.* University of Arizona Agricultural Experimental Station Technical Bulletin No. 68:181-222.

Nisbet, R.A. 1969. *Social Change and History: Aspects of the Western Theory of Development.* London: Oxford University Press.

Nuttall, L.J. 1878a. Letter to President John Taylor. Dated September 24, 1878, from Sunset, Arizona Territory. Historical Department, Church of Jesus Christ of Latter-Day Saints. Salt Lake City.

-----. 1878b. Letter to President John Taylor. Dated September 26, 1878, from Woodruff, Arizona Territory. Historical Department, Church of Jesus Christ of Latter-Day Saints. Salt Lake City.

-----. 1878c. Letter to President John Taylor. Dated September 29, 1878, from Clark's Ranch, Showlow near the Summit of the Mogollon Mountains, Arizona. Historical Department, Church of Jesus Christ of Latter-Day Saints. Salt Lake City.

O'Dea, T.F. 1957. *The Mormons.* Chicago: University of Chicago Press.

Odum, E.P. 1971. *Fundamentals of Ecology, Third Edition.* Philadelphia: Saunders.

Odum, H.T. 1971. *Environment, Power and Society.* New York: John Wiley and Sons.

Odum, H.T., and R.C. Pinkerton. 1955. Times Speed Regulator: the Optimum Efficiency for Maximum Output in Physical and Biological Systems. *American Scientist* 43:331-343.

Paine, R.T. 1966. Food Web Complexity and Species Diversity. The *American Naturalist* 100:65-75.

Patrick, R., M. Hohn, and J. Wallace. 1954. A New Method of Determining the Pattern of the Diatom Flora. *Notulae Natura* 259. Academy of Natural Sciences of Philadelphia.

Patten, B.C., and E.P. Odum. 1981. The Cybernetic Nature of Ecosystems. *American Naturalist* 118:886-895.

Peterson, C.S. 1967. Settlement on the Little Colorado 1873-1900: A Study of the Processes and institutions of Mormon Expansion. PhD. Dissertation, Department of History, University of Utah.

-----. 1970. "A Mighty Man Was Brother Lot": A Portrait of Lot Smith, Mormon Pioneer. *The Western Historical Quarterly* 1:393-414.

-----. 1973. *Take Up Your Mission: Mormon Colonizing along the Little Colorado River 1870-1900*. Tucson: University of Arizona Press.

-----. 1976. A Mormon Town: One Man's West. *Journal of Mormon History* 3:3-12.

Peterson, S. 1978. Shepherd of the Open Range. *Arizona Highways* 54 (August):2-9.

Pianka, E.R. 1966. Latitutinal Gradients in Species Diversity: A Review of Concepts. *The American Naturalist* 100:33-46.

Piddocke, S. 1965. The Potlatch System of the Southern Kwakiutl. *Southwestern Journal of Anthropology* 21:244-264.

Pielou, E.C. 1975. *Ecological Diversity*. New York: Wiley.

Plog, F.T. 1973. Diachronic Anthropology. In *Research and Theory in Current Archaeology*, edited by C. Redman, pp. ____. New York: John Wiley and Sons.

Popper, K. 1957. *The Poverty of Historicism*. New York: Harper and Row.

Porter, R.E. n.d.a. This Is My Own My Native Land. Box 13, Book 2. Special Collections, Northern Arizona University Library. Flagstaff.

-----. n.d.b. Joseph City Irrigation Company. Box 4, Folder 2. Special Collections, Northern Arizona University Library. Flagstaff.

-----. n.d.c. The Little Colorado River Valley: Its Description, Its History, Its Settlement by the Mormons. Historical Department, Church of Jesus Christ of Latter-Day Saints. Salt Lake City.

-----. n.d.d. Miscellaneous Writings. Box 3, Folder 8. Special Collections, Northern Arizona University Library. Flagstaff.

Rappaport, R. 1968. *Pigs for the Ancestors: Ritual in the Ecology of a New Guinea People*. New Haven: Yale University Press.

Rapport, D.J., and J.E. Turner. 1977. Economic Models in Ecology. *Science* 195:367-373.

Richards, J.M., and A.B. Westover. 1964. *Unflinching Courage, Joseph City, Arizona*. Joseph City, Arizona: John H. Miller.

Ricklefs, R.E. 1987. Community Diversity: Relative Roles of Local and Regional Processes. *Science* 235:167-171.

Rodhe, W. 1955. Can Plankton Production Proceed during Winter Darkness in Subarctic Lakes? *Proceedings of the International Association of Theoretical and Applied Limnologists* 12:117-122.

Rose, D. 1981. *Energy Transition and the Local Community: A Theory of Society Applied to Hazleton, Pennsylvania*. Philadelphia: University of Pennsylvania Press.

Rosenzweig, M.L. 1968. Net Primary Productivity of Terrestrial Communities: Predictions from Climatological Data. *American Naturalist* 102:67-74.

-----. 1976. On Continental Steady States of Species Diversity. In *Ecology and Evolution of Communities*, edited by M.L. Cody and M. Diamond, pp. 121-140. Cambridge: Belknap Press.

Russo, J. 1964. *The Kaibab North Deer Herd: Its History, Problems and Management*. State of Arizona Game and Fish Department, Bulletin 7.

Ruhle, E.E. 1973. Genetic and Cultural Pools: Some Suggestion for a Unified Theory of Bio-Cultural Evolution. *Human Ecology* 1:201-215.

Sahlins, M.D. 1958. *Social Stratification in Polynesia*. Seattle: University of Washington Press.

-----. 1961. The Segmentary Lineage: An Organization of Predatory Expansion. *American Anthropologist* 63:322-343.

-----. 1965. On the Sociology of Primitive Exchange. In *The Relevance of Models for Social Anthropology*, edited by M. Banton, pp. 139-236. London: Tavistock.
St. Johns Stake. n.d. Minutes of the St. Johns Stake Conferences. Historical Department, Church of Jesus Christ of Latter-Day Saints. Salt Lake City.
St. Joseph United Order. n.d.a. Minutes of the United Order Joseph City, Arizona, 1876-1887. Historical Department, Church of Jesus Christ of Latter-Day Saints. Salt Lake City.
-----. n.d.b. Articles of Association. Historical Department, Church of Jesus Christ of Latter-Day Saints. Salt Lake City.
Salt River Project. 1974. *Environmental Report, Arizona Station Project: Snowflake and St. Johns Generating Station Sites*. Phoenix: Salt River Project.
Samuelson, P. 1958. *Economics: An Introductory Analysis, Fourth Edition*. New York: McGraw-Hill.
Sanders, H.L. 1968. Marine Benthic Diversity: A Comparative Study. *The American Naturalist* 102:243-282.
Sanders, W.T., and D.L. Nichols. 1988. Ecological Theory and Cultural Evolution in the Valley of Oaxaca. *Current Anthropology* 29:33-80.
Sanders, W.T., and B.J. Price. 1968. *Mesoamerica: The Evolution of A Civilization*. New York: Random House.
Schmatz, E.M., et al. 1968. *Livestock-Poisoning Plants of Arizona*. Tucson: University of Arizona Press.
Schneider, H.K. 1974. *Economic Man: The Anthropology of Economics*. New York: Free Press.
Schoener, T.W. 1971. Large-Billed Insectivorous Birds: A Precipitous Diversity Gradient. *Condor* 73:154-161.
Service, E. 1971. *Primitive Social Organization: An Evolutionary Perspective, Second Edition*. New York: Random House.
Shannon, F.A. 1945. *The Farmer's Last Frontier, Agriculture, 1960-1897*. New York: Holt, Rinehart and Winston.
Simon, J.L. 1977. *The Economics of Population Growth*. Princeton: Princeton University Press.
Singer, H. 1950. The Distribution of Gains between Investing and Borrowing Countries. *American Economic Review*, Papers and Proceedings.
Slobodkin, L.B. 1968. Toward a Predictive Theory of Evolution. In *Population Biology and Evolution*, edited by R.C. Lewontin, pp. 187-205. Syracuse: Syracuse University Press.
Slobodkin, L.B., and H.L. Sanders. 1969. On the Contribution of Environmental Predictability to Species Diversity. *Diversity and Stability in Ecological Systems*. Brookhaven Symposia in Biology 22:82-95.
Smith, C.A. ed. 1976. *Regional Analysis, Volumes I and II*. New York: Academic Press.
Smith, E.A. 1984. Anthropology, Evolutionary Ecology, and the Explanatory Limitations of the Ecosystem Concept. In *The Ecosystem Concept in Anthropology*, edited by E. Moran, pp. 51-86. Boulder: Westview Press.
Smith, J.M. 1964. Group Selection and Kin Selection. *Nature* 201:1145-1147.
Smith, S. 1934. A Historical Survey of the Northeastern Section of Arizona, Its Settlement, and Development into Latter-Day Saint Stakes 1876-1934. M.A. Thesis, Department of History, Brigham Young University.

References Cited

Snowflake Stake. n.d. Minutes of the Snowflake Stake Conferences. Historical Department, Church of Jesus Christ of Latter-Day Saints. Salt Lake City.

Spencer, J.S., Jr. 1966. *Arizona's Forests*. U.S. Forest Service Resource Bulletin INT-6. Washington, D.C.: U.S. Department of Agriculture.

Spooner, B., ed. 1972. *Population Growth: Anthropological Implications*. Cambridge: M.I.T. Press.

Stebbins, G.L. 1968. Integration of Development and Evolutionary Progress. In *Population Biology and Evolution*, edited by R.C. Lewontin, pp 17-36. Syracuse: Syracuse University Press.

Stegner, W. 1942. *Mormon Country*. New York: Hawthorne Books.

-----. 1964. *The Gathering of Zion: The Story of the Mormon Trail*. New York: McGraw-Hill.

Steward, J. 1955. *Theory of Culture Change: The Methodology of Multilinear Evolution*. Urbana: University of Illinois Press.

Struever, S., and G.L. Houart. 1972. An Analysis of the Hopewell Interaction Sphere. In *Social Exchange and Interaction*, edited by E.N. Wilmsen, pp. 47-79. Ann Arbor: University of Michigan Press.

Stubblefield, T.M. 1953. *Economic Survey of Navajo County*. Agricultural Extension Service, University of Arizona. Tucson.

Tanaka, J. 1976. Subsistence Ecology of Central Kalihari San. In *Kalihari Hunter Gatherers: Studies of the !Kung San and Their Neighbors*, edited by R.B. Lee and I. DeVore, pp. 98-119. Cambridge: Harvard University Press.

Tanner, G.M., and J.M. Richards. 1977. *Colonization on the Little Colorado: The Joseph City Region*. Flagstaff, Arizona: Northland Press.

Terborgh, J. 1971. Distribution on Environmental Gradients: Theory and a Preliminary Interpretation of Distributional Patterns in the Avifauna of the Cordillera Vilcabama, Peru. *Ecology* 52:23-40.

Thomas, E.N. 1960. Some Comments on the Functional Bases for Small Iowa Towns. *Iowa Business Digest* 31:10-16.

Tylor, E.B. 1871. *Primitive Culture*. London: John Murray.

Udall, D.K. n.d. David K. Udall Journal. Historical Department, Church of Jesus Christ of Latter-Day Saints. Salt Lake City.

Underwood, A.H. 1970. A Study of Ranch Management Practices in Navajo County, Arizona. M.A. Thesis, Department of Agricultural Education, University of Arizona.

U.S. Department of Interior. 1946. *The Colorado River*. Washington, D.C.: U.S. Government Printing Office.

Vandermeer, J.H. 1972. Niche Theory. *Annual Review of Ecology and Systematics* 3:107-132.

Vayda, A., and B.J. McCay. 1975. New Directions in Ecology and Ecological Anthropology. *Annual Review of Anthropology* 4:297-306.

Vayda, A., and R. Rappaport. 1968 Ecology: Cultural and Non-Cultural. In *Introduction to Cultural Anthropology*, edited by J.A. Clifton, pp. 477-497. New York: Houghton-Mifflin.

Vogt, E.Z. 1955. *Modern Homesteaders: The Life of a Twentieth-Century Frontier Community*. Cambridge: Belknap Press.

Vogt, E.Z., and E.M. Albert, eds. 1970. *People of Rimrock: A Study of Values in Five Cultures.* New York: Atheneum.

Vogt, E., and T. O'Dea. 1953. A Comparative Study of the Role of Values in Social Action in Two Southwestern Communities. *American Sociological Review* 18: 645-654.

VonBertalanffy, L. 1968. *General Systems Theory.* New York: Braziller.

Waddington, C.H. 1957. *The Strategy of the Genes.* London: Allen and Unwin.

Wallerstein, I. 1974. *The Modern World-System.* New York: Academic Press.

Warner, R..W. 1968. United Order of Little Colorado Stake in Arizona. Unpublished manuscript. Historical Department, Church of Jesus Christ of Latter-Day Saints. Salt Lake City.

Wayte, H.C. 1962. A History of Holbrook and the Little Colorado Country (1540-1962). M.A. Thesis, Department of History, University of Arizona.

Weather Bureau. n.d. *Climatological Data, Arizona Section.* Washington, D.C.: U.S. Department of Agriculture.

Webb, W.P. 1931. *The Great Plains.* New York: Grossett and Dunlap.

White, B. 1973. Demand for Labor and Population Growth in Colonial Java. *Human Ecology* 1:217-236.

-----. 1967. The Economic Importance of Children in a Javanese Village. In *Population and Social Organization*, edited by M. Nag, pp. 127-146. The Hague: Mouton.

White, L.A. 1959. *The Evolution of Culture.* New York: McGraw-Hill.

Whittaker, R.H. 1975. *Communities and Ecosystems, Second Edition.* New York: Macmillan.

Whittaker, R.H., and G.M. Woodwell. 1972. Evolution of Natural Communities. In *Ecosystem Structure and Function*, edited by J.A. Weins, pp. 137-159. Eugene: University of Oregon Press.

Wilkinson, R. 1973. *Poverty and Progress: An Ecological Perspective on Economic Development.* New York: Praeger.

Williams, G.C., ed. 1971. *Group Selection.* Chicago: Aldine-Atherton.

Wilson, D.S. 1980. *The Natural Selection of Populations and Communities.* Menlo Park: Cummings.

Wilson, E.O. 1968. The Ergonomics of Caste in the Social Insects. *American Naturalist* 102:41-66.

-----. 1971. *The Insect Societies.* Cambridge: Belknap Press.

Winter, S.G., Jr. 1964. Economic "Natural Selection" and the Theory of the Firm. *Yale Economic Essays* 4:225-272.

Winterhalder, B., and E. Smith, eds. 1981. *Hunter-Gatherer Foraging Strategies: Ethnographic and Archaeological Analyses.* Chicago: University of Chicago Press.

Woodruff Irrigation and Recreation Project. n.d. Woodruff Irrigation and Recreation Program. Woodruff, Arizona.

Young, G., and C.A. Broussard. 1986. The Species Problem in Human Ecology. In *Human Ecology: A Gathering of Perspectives*, edited by R.J. Borden, pp.55-67. College Park, Maryland: Society for Human Ecology.

INDEX

Alpine, Arizona, 21, 32, 34, 36, 46, 52, 55, 74, 84, 88, 89, 92, 114, 119, 120, 122, 123, 134, 187
American Colonization Company, 20, 47, 187
Anti-Bigamy Act, 174
anti-Mormon activities, 47, 171, 175, 177, 179, 186
Apache County, Arizona 34, 52, 109, 112, 171, 174, 179, 183-186
Arizona Cooperative Mercantile Institution (ACMI), 128, 136, 152-155, 160-161, 167, 175, 179, 198
Atlantic and Pacific Railroad, 1, 2, 3, 6, 16, 32, 34, 46, 125, 139, 159, 160, 161, 162, 164-168, 170, 171, 176-184, 198, 199, 203
Aztec Land and Cattle Company, 6, 109, 111, 119, 164, 167-171, 177, 178, 180-184, 186, 187
Barth, Solomon, 31, 137, 184
Brigham City, Arizona, 21, 23, 24, 26-29, 31, 32, 34, 48-52, 135, 143, 144, 147, 148, 158, 177, 180
cattle, 6, 31, 34, 35, 49, 51, 53, 89, 109, 111, 112, 119, 144, 158, 164, 167-171, 180-183, 186, 187
chaining, 111, 119
community diversity: in ecological communities, 56-58, 62-64; in human communities, 66-67; defined among Little Colorado settlements, 73
community productivity: in ecological communities, 64,66; in human communities, 67-68, 71, 72, 77
Concho, Arizona, 21, 46, 49
conjoint enterprises, 3, 6, 8-10, 135, 141, 143, 148, 155, 156, 196-198, 200
Continental Cattle Company, 181, 182
control circuits: in ecological systems, 62-63; in human communities, 67, 70
Daggs Dam and Reservoir, 94, 104, 107-109, 123, 126, 132, 133, 139
dam failures, 2, 5-7, 9, 16, 115, 116, 122, 123, 128, 130, 131, 133-136, 142, 145, 146, 155, 191, 193, 194, 202
Eagar, Arizona, 21, 32, 36, 46, 52, 74, 90-92, 100, 115, 119, 123, 134, 156, 186, 187, 193, 194
Eastern Arizona Stake, 17, 52, 123, 147, 153, 156, 160, 169, 170, 175, 179
Edmunds Act, 174
energy drains: in ecological communities, 58-59; in human communities, 68-70
energy subsidies: in ecological communities, 58-59; in human communities, 68-70
environmental stability/instability: in ecological systems, 3, 33, 60-63; role in human communities, 68-70; among Little Colorado settlements,73-74, 148, 156, 191, 197-200, 202
environmental productivity:in ecological systems, 59-60; role in human communities, 68-69
environmental regulation, 9, 72, 155-157, 197, 198, 200, 201
Flake Ranch Reservoir, 132
Flake, William, 31, 51, 132, 139, 143
Forest Dale, Arizona, 30, 51
Fort Apache, 50, 51, 179
freighting, 50, 145, 179
Greer, Arizona, 32
gristmill, 142, 157
groundwater, 93, 117,
growing season, 2, 25, 32, 53, 69, 73, 84, 91, 93, 101, 114-116, 120, 122, 135, 190, 191, 193, 194

Hashknife outfit, 168, 169, 177, 181-183, 186
highland settlements, 120, 123, 134, 154, 190, 191, 194, 196, 203
Hispanic population, 21, 130, 137, 171-173, 183-185, 203
Holbrook, Arizona, 20, 46, 79, 83, 84, 94, 95, 100-103, 117, 118, 122, 138, 153, 160, 164-167, 169, 173, 178-180, 183, 194, 196, 202
Holbrook Anticline, 117, 194
Hunt, Arizona, 129, 138
intermediate settlements, 123 156, 193-197, 199
irrigation 2, 5, 16, 23, 24, 48, 51, 54, 62, 65, 68, 73, 83, 90-93, 95, 101, 108, 112-118, 121-123, 125, 129-139, 144, 148, 151, 159, 170, 172, 176, 184, 190-192, 194
Kanab, Utah, 23, 25, 164, 180
labor on dams,
man-days, 23, 27, 29, 48, 49, 124, 127, 136, 142, 144, 145, 158, 159
team-days, 23, 29, 124, 127, 136
Leone, Mark, 3, 5-9, 12, 17, 56, 76, 118, 122, 123, 131, 135, 150, 153, 159, 179, 186
Lightfoot, Kent, 3, 7-9
Little Colorado River, 1-4, 7-13, 15-17, 19-21, 24, 25, 27, 29, 31, 33, 35, 46, 47, 49-52, 55, 70-74, 79, 81, 83, 84, 86, 88-90, 92, 94, 95, 100-104, 107-109, 112-119, 121-123, 127, 129, 130, 135-137, 141, 142, 147, 148, 150, 152, 155, 157-160, 161, 162, 164, 167-169, 171, 175-181, 183-187, 189-191, 195, 196, 199-202
Little Colorado Stake, 8, 17, 28-34, 49, 52, 123, 124, 127, 171, 186
livestock, 31, 36, 52, 53, 86, 89, 111, 112, 138, 143, 158, 164-166, 168, 171, 178, 180, 183, 187
livestock productivity, 112, 178, 187
Lone Pine Dam, 139
lower valley settlements, 8, 9, 48, 52, 115, 121, 123, 130, 135, 142, 144, 146-148, 191, 192, 195-197, 199, 200

Lyman Dam, 94, 95, 118, 127, 129, 131, 135, 137-139
Mormon Church, 1-3, 5, 7, 8, 16, 17, 19, 20, 27, 30, 31, 34, 46-48, 50, 51, 54, 65, 71, 73, 78, 127-129, 131, 134, 136, 137, 140, 141, 142, 144, 148, 150-154, 157, 159, 160, 161, 162, 167, 170, 171, 174-179, 184, 186, 187, 194, 198-200
Mormon Dairy, 49, 142-144, 157, 183
Navajo County, Arizona, 52, 184, 186
niche concept, 55-56; in human communities, 64-66
non-Mormons, 1-3, 16, 19, 20, 30, 46, 52, 152, 155, 161, 167, 171, 172, 177, 185-187, 194
Nutrioso, Arizona, 32, 119, 120
Obed, Arizona, 21, 23-27, 48, 49, 123, 135, 158, 177, 180
Obed Meadow, 49
Old Taylor, Arizona, 30, 31, 50, 135, 177
overgrazing, 108, 111, 112, 115, 116, 119, 182, 183, 192
Peterson, Charles S., 3-5, 7, 9, 10, 20, 21, 23, 24, 27, 30-32, 34, 47-49, 51, 119, 122, 127-130, 133, 134, 142, 143, 150-154, 158, 162, 164, 171-173, 175, 176, 179, 182-184, 186, 202, 203
Pleasant Valley War, 169, 183
polygamy, 3, 154, 171, 174, 175
polygamy raids, 154, 175
precipitation, 15, 27, 34, 53, 68, 81, 83, 84, 86, 88, 89, 91, 93-95, 104, 111, 115-117, 121, 190
range deterioration, 109, 111, 119, 169, 171, 177, 183
range productivity, 111
redundancy: in ecological systems, 63; in human communities, 71
regulation: in ecological systems, 62-63; in human environmental relations, 70-71
Round Valley, 21, 32, 52, 113, 119, 177, 186
Round Valley Water Users Association, 140

Index

St. Johns, Arizona, 17, 21, 28, 31, 32, 34, 35, 46, 49, 52, 54, 81, 83, 84, 86, 88, 91, 92, 94, 95, 101, 104, 107, 111, 113, 115, 117, 118, 123, 125-127, 129-135, 137-139, 154, 156, 162, 171-175, 177, 184-186, 193, 194, 202, 203
St. Johns Irrigation Company, 113, 130, 137, 138
St. Johns Ring, 172-175, 185, 186
St. Johns Stake, 34, 35, 52, 123, 131, 154
St. Joseph, Arizona, 6, 21, 23-36, 46-50, 52, 74, 86, 90, 92, 113, 115, 121-127, 129-136, 138, 144-148, 151, 158, 161, 164, 165, 167, 177-179, 184, 187, 192, 196, 202, 203
sand dams, 24, 48, 126
sawmill, 142-145, 157
sheep, 27, 89, 109, 112, 119, 145, 164, 169, 180, 183
sheepherding, 137, 172
Showlow, Arizona, 21, 30, 36, 46, 49, 52, 74, 79, 83, 90, 113, 114, 116, 119, 121, 123, 129, 134, 180
Shumway, Arizona, 104, 113, 118, 139, 167, 180
Silver Creek, 1, 21, 25, 27, 29-31, 34, 49, 51, 91, 94, 95, 100, 102-104, 107-109, 113, 115, 117, 121, 122, 128, 129, 132, 133, 136, 137, 139, 180, 194
silt, 5, 83, 89, 90, 102, 109, 112-115, 118, 121, 135, 148
Slough Dam, 131, 137
Snowflake, Arizona, 17, 21, 25, 27, 29-31, 34, 49, 51, 79, 83, 84, 86, 88, 91, 92, 103, 104, 107, 108, 111, 113, 115-117, 119, 121, 122, 123, 125, 126, 129, 132-139, 143, 147, 154, 156, 160, 161, 162, 167, 169, 170, 175, 177, 180, 183, 184, 193-196, 198, 199, 202, 203
Snowflake and Taylor Irrigation Company, 132, 139
Snowflake Stake, 52, 123, 154
soil quality, 2, 15, 73, 89-92
Springerville, Arizona, 21, 32, 46, 49, 52, 81, 83, 100, 101, 121, 122, 171, 186
Stinson's ranch, 27, 31, 49
Sunset, Arizona, 21, 23-29, 31, 32, 34, 48-50, 52, 135, 142-144, 146-148, 158, 177, 180
tannery, 142, 144, 157
Taylor, Arizona, 21, 30, 31, 34, 36, 46, 50-52, 91, 103, 104, 108, 113, 115, 117-119, 123, 129, 132-136, 139, 156, 170, 174, 177, 180, 184, 193-196, 202, 203
tithing: totals collected, 36, 53, 73, 74, 150, 202; per capita, 36, 73, 74, 114, 202; redistribution, 3, 5-9, 12, 16, 46, 75, 135, 148, 151, 153-157, 159, 167, 197, 198, 200, 201
United Order, 8, 25, 47, 48, 126, 135, 141-143, 150-152, 155, 158
Utah, 1, 19, 20, 23-25, 30-32, 36, 47, 51, 52, 86, 116, 145, 146, 159, 160, 162, 164, 174, 176, 179, 180, 203
Utah cattle, 31, 51
water: groundwater, 137; water quality, 118, 90, 102, 103, 107, 108, 126; surface water, 15, 69, 86, 93, 104, 112-115, 117, 118, 121, 137, 191, 193, 194
Winslow, Arizona, 47, 52, 83, 103, 119, 121, 165-167, 180, 181
Woodruff, Arizona, 6, 21, 27-31, 34, 36, 46, 49-51, 86, 90, 92, 95, 103, 104, 107-109, 113, 115, 118, 121-123, 127-134, 136-138, 148, 151, 153, 160, 161, 170, 177, 179, 184, 187, 192
Woodruff Irrigation and Recreation Project, 129, 136, 137
wool, 164-166
Young, John W., 12, 20, 50, 54, 109, 120, 162, 179, 187, 203